# The Troubled Waters of the Amazon

# The Troubled Waters of the Amazon

*The Plight of the Colombian Indians in Amazonia*

Verónica de Osa

ROBERT HALE · LONDON

© *Verónica de Osa 1990*
*First published in Great Britain 1990*

Robert Hale Limited
Clerkenwell House
Clerkenwell Green
London EC1R 0HT


British Library Cataloguing in Publication Data

Osa, Verónica de
The troubled waters of the Amazon: the plight of the
Colombian Indian in Amazonia.
1. Colombia. South American Indians. Social life
I. Title
986.1'00498

ISBN 0–7090–3930–1

Photoset in Palatino by
Derek Doyle & Associates, Mold, Clwyd.
Printed in Great Britain by
St Edmundsbury Press Ltd, Bury St Edmunds, Suffolk.
Bound by WBC Bookbinders Limited.

# Contents

BOGOTA

CAQUETA

YARI

PUTUMAYO

CAQUETA

Los Nenos

Santande

El Ultimo Retiro

CERROS DE
MAINE-HANARI

Chorrera

COLOM

Cahe

Caraparaná

Tidaima

Menave

Igara Paraná

San José

San Rafael

Sabalayaco

N

La Esperanza

0    40    80
Km

PU

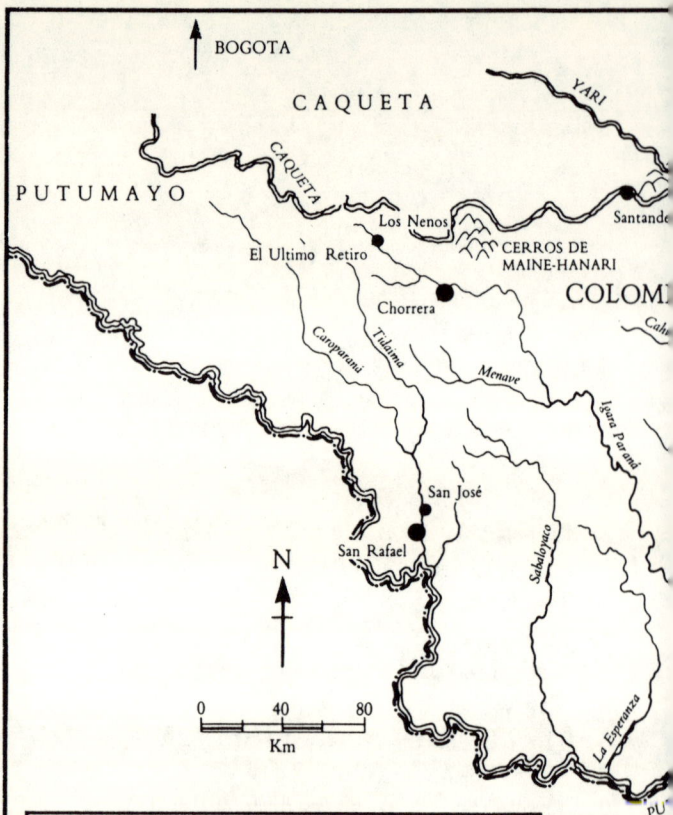

## SPECIAL DISTRICT
## OF AMAZONAS
### Hydrographical River Basins
### of Amazonas, Putumayo and Caquetá

| DISTANCES | | AIR | RIVER |
|---|---|---|---|
| LETICIA | Bogota | 1.085 KM | — |
| " | Pto. Nariño | 65 " | 84 KM |
| " | Tarapaca | 147 " | 642 " |
| " | Pedrera | 322 " | 1.731 " |
| " | Miriti | 417 " | 2.206 " |
| " | San Rafael | 465 " | 1.446 " |
| " | Chorrera | 507 " | 1.563 " |
| | | | |
| BOGOTA—LETICIA | via | FLORENCIA | 2.968 KM |
| BOGOTA—LETICIA | " | PASTO | 3.449 " |

VAUPES
A DEL
CUARA
APAPORIS
Popeyacá
NESA DEL SICAYARI
Yapurá
Japurí
Uamia
Ira
Guacaydco
Miriti
Comeyaco
Tunaré
Miriti Parona
Jaru
Negra
Caeni
CERROS DE
PEDRERA
Palma
Cahuinari
Yacare
Pedrera
La Tanira
Aupa
Arapa
AMAZONAS
COLINA DE FUTAHY
Pupuña
Pure
Arica
Tagna
Porvenir
Derecho
Sta Clora
BRAZIL
PERU
Tarapaca
Cotuhe
Loreto yaco
Boia Bossu
Atacuari
AMAZONAS
Pto.
Narino
San Martín
Mocagua
Atacuaro
Santa Sofia
Arora
Nezaret
Rondo
LETICIA

# Apostolic Prefecture of Leticia

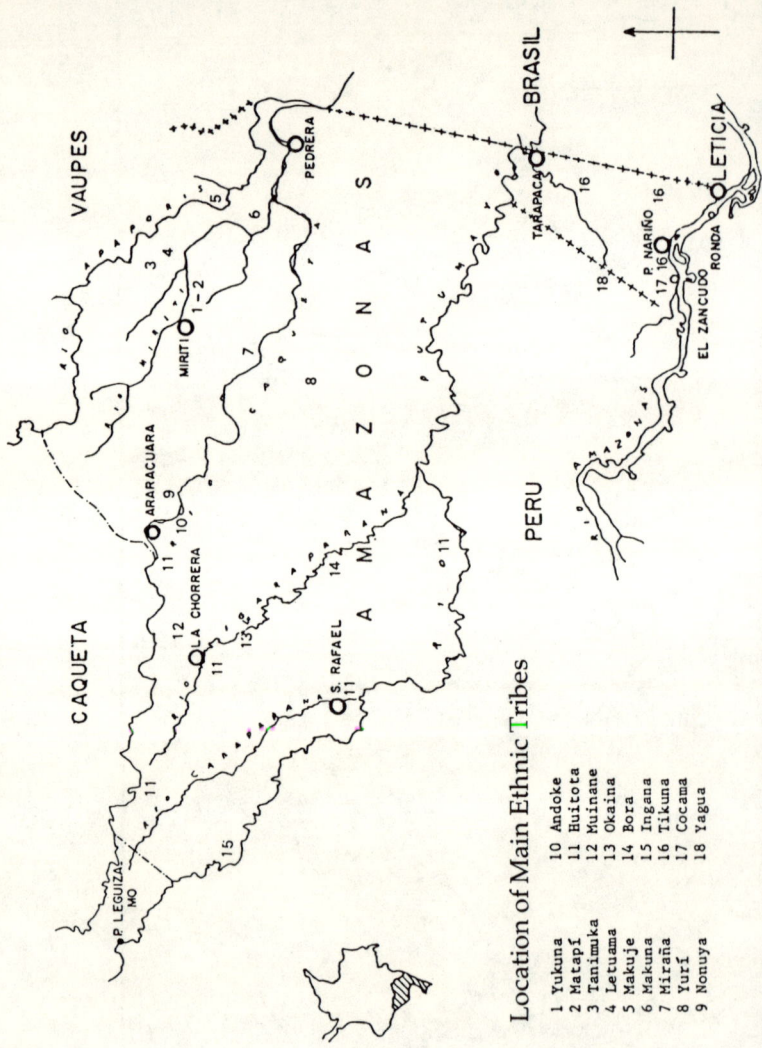

## Location of Main Ethnic Tribes

1 Yukuna
2 Matapí
3 Tanimuka
4 Letuama
5 Makuje
6 Makuna
7 Miraña
8 Yurí
9 Nonuya

10 Andoke
11 Huitota
12 Muinane
13 Okaina
14 Bora
15 Ingana
16 Tikuna
17 Cocama
18 Yagua

# Abbreviations

ACPO — Acción Cultural Popular (People's Cultural Action)

CILEAC — Centro de Investigaciones Linguisticas y Etnograficas de la Amazonia Colombiana (Centre for Indigene Linguistic Studies)

ICA — Instituto Colombiano Agropecuario (Colombian Institute for Agricultural and Cattle-raising Research, and its application)

ICETEX — Colombian Institute for Educational Credit of Technical Studies in the Exterior

INCORA — Instituto Colombiano de la Reforma Agraria (Colombian Institute for Agrarian Reform)

INDERENA — Instituto de Defensa de Recursos Naturales Renovables (Institute of Naturally Renewable Resources, autonomous, with juridical power and financial independence)

# 1   Gold and Blood

The natural resources of the part of Amazonia belonging to Colombia make it potentially one of the wealthiest areas in the world; yet it has remained one of the poorest.

The Spaniards who discovered the country in 1492 found it occupied by Indians, most of whom were hunters or nomad farmers. Having no domestic animals, they could only work the earth by hand. African Negro slaves were brought in by the Spaniards for the hard labour which the Indians could not handle satisfactorily.

The Spaniards brought to Amazonia their language, their Roman Catholic religion, their legal system. They brought their strength and their weaknesses, as did all colonizers. Spain conquered the land, and colonized it until 1810.

The original discovery had revealed vast numbers of Indians scattered all over Colombia, and said to be of Asiatic extraction. They can be classified as Andean, Caribbean and Plains Indians.

The first Spaniards came partly to convert the Indians to Catholicism, and partly from a sense of adventure. Trouble started when some of them found gold in the Antioquia region. Almost as soon as regular shipments of gold began to be despatched from 'Cartagena of the Indies' to the Spanish kings, the Caribbean ports were pestered with pirates. The Spaniards had seen works of art in gold made by the Indians, and they had found gold in the sands of Colombian rivers. Avarice had soon blinded them to the anguish of the Indians to whom it rightfully belonged.

Gold has been the stumbling-block of the entire history of colonization.

All through Colombia's history, the Church not only brought the Christian religion to the Indians but spread the ideas and institutions of Western civilization. It was responsible for the establishment and maintenance of almost all the schools in existence during the colonial period. Franciscans, Dominicans, members of the Order of Mercy and, later on, Jesuits and Augustinians, all participated actively in the colonial history. The Franciscans established convents in Cali, Popayán, in Cartagena and in Vélez, the Dominicans in Santa Fé (Bogotá), Pamplona and Popayán. In 1563 the Dominicans established the first Chair of Grammar and, some time later, that of Philosophy. The Jesuits came to Santa Fé in 1590, and nine years later they opened a college.

Earlier still, America had scarcely been discovered when national ambitions first became manifest.

Spain and Portugal appealed to the Vatican to divide the New World between them. Pope Alexander VI gave his judgement on the new continent on 4 May 1493. The Bull is called the *Alejandrina*, and the division seems simple enough. The Pope traced an imaginary meridian line beyond the Azores in the Atlantic Ocean, dividing territories to the right and to the left, settling thereafter the fight between Spain and Portugal.

Alexander VI's personal position is controversial. Unanimously elected as Pope, although he was the father of several offspring from his legitimate wife, his decision, called wise, has been legally contested. Was Pope Alexander VI entitled to give to the kings of Castile and Portugal dominion and direct sovereignty over the Indians? All the Pope could legally grant was the exclusive right to preach in the discovered territories, as well as those political and commercial benefits which were derived from the protection and defence of the faith in the new world. In truth, the fight was against the threatening advance of Islam along the Mediterranean and the Danube, which opposed Christianity, and even invaded the old kingdoms in India. In those times, the battle was not to be won by spiritual messages from the Pope; although it had been carried in the beginning by tolerant, unaggressive monks, it now required occupation of the intermediate territories, armed forces and the support of the Crown's treasury. The two Bulls of Alexander VI expressed prosecution of the same missionary ideal which it defended in accordance with the historical necessities of the passing centuries.

From this papal division of the newly conquered land has derived the very strong influence of the Church in South America. Orders had been given to the *conquistadores* to convert these new people to Christianity, and this was done, as was mentioned in Isabel the Catholic's testament. In 1554 a royal decree from Spain ordered the creation of schools for the indigenes. Since then the Spanish clergy have extended their influence over the entire colony. The Minister of Finance, the conservative Mosquera, consecrated the triumphs of the Colombian clergy by a concordat most advantageous to the Holy See. Modification of the concordat was introduced in 1842. Since then there have been bloody battles between the political parties, and between Catholics and Protestants accused of supporting the liberals.

Colombia is the only country in South America that is still bound by a concordat with Rome. So far it has been renewed about every twenty-five years, for it was considered that the work accomplished by missions in many regions of this very difficult country was on the whole laudable and irreplaceable. Men and women driven by a religious conviction will accomplish more for others than he who works merely from a mercenary motivation.

The pioneer who achieved most for the slaves was St Pedro Clavel, born in 1580 in Verdú, Catalonia. A Jesuit father, he embarked with other members of the Society of Jesus in 1610, became a priest in 1616, then returned to Cartagena, where he died on 3 September 1654. He has so far remained the only canonized Colombian saint and is considered the patron of the Negro missions.

In recent times Camilo Torres was the first Catholic priest in South America to force the Catholic Church to become more socialistic. It was 'the Church of the Conquerors', and Camilo Torres felt it should become 'the Church of the Poor'.

# 2 Defence of a Mission

Pope Alexander VI laid down a condition with his division of the New World between Spain and Portugal that missionaries should be sent to South America – 'upright and God-fearing men of learning' – to instruct the native peoples in the Catholic faith. The Spanish monarchs accepted the delegation willingly. In her famous last will, Queen Isabel wrote that, 'Our principal intention was to convert the peoples of the Indies to our Holy Catholic Faith.' Charles V in 1532 expressed his gratitude to God, 'from whom we were given grace to discover the Indies'. Later the Crown received the 'royal patronage' from the Pope, making the Spanish kings his vicars in America.

This was the foundation laid for the evangelization of the whole of the American Indies (South America) by the Church of Spain, an extraordinary religious achievement.

The arrival of the first missionary on Amazon territory was in 1542. Brother Diego de Carvajal took part in the expedition of Francisco de Orellana, the *conquistador* wounded by the legendary Amazonas during their meetings at the River Cundurises. The first missionaries in the region were Franciscans, but there is little information about the colonial period of the south-eastern part of Colombia, except that Brazilian-Portuguese slave-hunters had already deeply modified the traditional structures of the indigenous population as early as the second half of the fifteenth century.

The Indians were to be grouped and transferred, but they rebelled and killed some Franciscan missionaries. The first missionary centres were not permanent, because there was a great lack of human and economic resources. The Indians tried to flee, they suffered from disease and were attacked by other tribes. Finally the missionaries were obliged to give up their centre on the River Caquetá, and their situation had become really critical by the beginning of 1790. Before this, La Condamine, the French traveller who navigated the Amazon river as early as 1743, had reported the permanent presence of Portuguese-Brazilian slave-hunters in the area.

Later on, von Humboldt and Bonpland personally observed the participation of some Indian chiefs in the slave trade. In the middle of the eighteenth century, the Marquess of Pombal reorganized the relationship between settlers, missionaries and Indians. The ecclesiastical powers were abolished and the civil employees had full control over the Indians. These Portuguese missionaries were given only the simple status of priests.

The earliest presence of the Capuchin Franciscan fathers from Catalonia in Latin America dates from the year 1681 in Trinidad and from their Orinoco Convoy in 1687. Before this, in 1612, the French Capuchins had founded a residence in Maranhao, and Friar José de Tremblay, the *eminence grise*, had settled briefly in Quebec. The Capuchins from Andalusia, Spain, founded a mission in Uraba in 1647, which was abandoned when they were recalled to Europe. It was they who built the first South American Capuchin monastery, '*el convento del Socorro*', whose first father superior, appointed in 1895, was Agostín del Alcoy. The presence of the Catalan Capuchin fathers dates from the end of the nineteenth century, when coming from Trinidad and Ecuador, they took charge of the south-east of Colombia. The Capuchin Amazon missionaries (who are still working in the Amazon part of Colombia) were certainly obedient to their calling and to Christ's words: 'Go into all the world and preach and publish openly the good news to every creature' (Mark 16:17).

Although they knew how the Franciscan missionaries had been slaughtered, these Capuchin fathers left their homeland for the then inhospitable Americas, achieving lasting memorials to their faith, founding great cities, such as San Francisco in the United States, and Florencia in Colombia, and in other places in Latin America.

In the greater part of Colombia, the Indians suffered the rule of the *encomenderos*, Spanish settlers who shared out the Indians among themselves in order to exercise control over them. Towards the south-east, the *encomienda*\* extended only to the valley of Sibundoy, while the vast region of the Amazon it proved impossible to incorporate the Indians into the colonial economy. Then the Colombian part of Amazonia was only partially accessible by road, where air travel has now largely replaced water transport, formerly the only means of travel. The mighty Amazon river itself, whose secrets have never been entirely discovered, is the most important of these waterways. Of its 6,420 kilometres, only 120 belong to Colombia, forming the frontier with Peru and Brazil. This is – since the loss of Panama – Colombia's only direct access to the Pacific Ocean.

It was not until the end of the nineteenth century, with the foundation of the Peruvian company Casa Arana, that the Indians of the Amazon were really enslaved. Three hundred years earlier, a Jesuit priest, Father Ferrer, in his *Convoy del Indio*, had worked as one of the earliest

---

\* *encomienda*: estate granted by the Spanish kings

missionaries among cannibal tribes. It is believed that he was one of the first Jesuit martyrs: he disappeared and the only traces of him which could be found were his spectacles!

In 1717, by Spanish royal decree, the viceroyalty of New Granada was founded in the south-eastern part of Colombia. It comprised the provinces of Santa Fé, Cartagena, Santa Marta, Maracaibo, Caracas, Antioquia, Guayana and Popayán, all under the presidency of Quito. In 1740 the borders between New Granada and Peru were established, bringing the northern region of the Amazon into New Granada. This was confirmed in 1770 by the treaty of San Ildefonso between Spain and Portugal. However, during the period of the Republic, the south-east of the country remained forgotten by its own government. The economic problems of the newly liberated Colombia were quite different, as indeed they are today, from those of the Amazon. Of all the 219 paragraphs of the Constitution of Colombia of 1832, none was devoted to the territories of the Amazon. Not until the charter of 1835 was it stated that all the lands of the Guajira, the Caquetá and other lands in the south of the country, '... uninhabited by civilized peoples, should be administered and governed by means of a separate legislation'. In the Constitution of 1843, these territories, far away from the capital and sparsely populated, are referred to as '*Territorios Nacionales*'.

There were no navigation treaties between Colombia and Brazil until General Rafael Reyes (later President of Colombia) made a personal application on behalf of the Colombian firm Elias Reyes to import and export goods on Colombian ships, between the ports on the Amazon and the interior of Colombia, via the Putumayo river. In 1876 the firm's ship the *Tundama* became the first steamer to navigate the Putumayo. No duty was paid on transit through Brazilian waters the entire time that the company was in operation, until 1884.

General Obanda had found slave-traders in the lower Putumayo. In 1841 he uncovered the existence of an intense trade between Pasto and that region. All of this was confirmed by General Reyes several years afterwards: 'In our first trips we met, in the lower part of the rivers Putumayo and Caquetá, big ships with sails and oars serving the horrible trade of selling Indians of both sexes, just so as to augment the war between the different tribes who are enemies, and will be transported to the Amazon river to be sold as slaves. On one occasion, we had to combat with some fifty naked, wild Indians, starved with hunger, who were ailing at the bottom of the ship. We took these dealers of human flesh as prisoners and brought them to the Brazilian authorities, giving the poor Indians their freedom with their own tribes.' This was the situation which the Capuchin fathers found when they arrived as missionaries.

In 1896, following an appeal made by Monsignor Caycedo, Bishop of Pasto, Capuchin missionaries arrived in the region to attend to its spiritual welfare. The missionaries came to Mocoa, in the east of the Caquetá, for a trial period of five years. After this they decided to settle in Mocoa, and in 1904 the Apostolic Prefecture of the Caquetá was founded by papal decree. In 1906 Father Fidel de Montclar was designated Apostolic Prefect.

At that time there were no roads into the interior of the country. In order to reach Mocoa from Pasto (the nearest largest town), the traveller had to hack his way through jungle, running the risk of falling into deep pits hidden in the undergrowth. Those natives who had not benefited from the civilizing influence of the European missionaries could also pose a threat. The other non-Indian inhabitants of the region, with whom the natives came into contact, were Peruvians, who had no inclination whatsoever to educate the Indians: their only motive for being in the

Amazon was greed, their only purpose to exploit the indigenous population in pursuit of the profits to be gained by exporting rubber or quinine.

The Capuchin missionaries of 1896 came to a country which nominally formed part of Colombia but which in reality was ruled by Peruvian nationals. Dr Enrique Puertas, a special commissioner of the Putumayo, wrote: 'Without the missionaries, the road between Pasto and Umbrai would not have been built, and it would still not be possible to exploit the fertile land of the Sibundoy Valley; Sucre would not exist, nor would San Francisco, Umbria, Puerto Asis or other important settlements in the lower Putumayo. The road has been called a "deed of charity"; it allows the traveller to reach Umbria and beyond, right into the Putumayo.'

The creation of the Apostolic Prefecture increased the number of missionaries in the region and intensified efforts to convert the Indians to Christianity.

The Capuchin fathers founded the villages of Puerto Asis, Umbria, Alvernia, Sucre, Belen and others. They would select an area suitable for the establishment of a settlement and supervise construction work, so that the houses were built close together, instead of being scattered among the fields as was the normal practice. The pattern of the villages can be traced back to that of the city of Florencia, designed by the Capuchin Father Doroteo. A reporter from Colombia's daily paper *El Tiempo* wrote: 'On my outward journey, I met a Capuchin monk measuring the land, kneeling deep in mud; when I returned, I found the village of Sucre.'

The missionaries began to build villages and settlements in the south. They started near Pasto, in the valley of Sibundoy, where today, in a botanical garden sponsored by the World Wildlife Fund, are cultivated medicinal herbs traditionally grown by the local Indians. At that time over 5,000 Indians lived in the valley.

From Pasto the missionaries advanced to Florencia and then to San Vicente. They were unable to penetrate further than Caucayá because the whole of the Caraparaná, with its waterfalls, and part of the Igaranaparaná, which marked the limit of the Apostolic Prefecture, came under the rule of Peru, through the infamous Casa Arana, as will be explained later on. This situation persisted until 1932, when the Holy See, with the agreement of the Colombian government, was able to extend the official boundaries of the Prefecture. It then went right up the Amazon, as far as the Brazilian village called Tabatinga, and from there to the point where the Caquetá river meets the Apopories.

From the beginning, the Capuchin missionaries placed the utmost importance on educating the Indians. They considered education essential as a basis for conversion to Christianity and extended this principle to the local white population.

The number of schools founded by the Capuchin fathers grew apace: in 1906 there were eleven, with 493 pupils in all; by 1933 the number had increased to sixty-five schools and 2,700 pupils.

Father Gaspar Mignell Moncanill, author of *Los Capuchinos 1906–1933*, wrote that, 'The efforts demanded of the missionaries to keep this number of pupils in schools are quite unbelievable; they have to strive constantly to persuade the pupils' parents, not only the Indians but also the "civilized" inhabitants, to bring their children to school. Thus in the village of Santiago there are 420 boys and girls who attend school regularly, with only a very few remaining who do not receive education In Las Casas, on the other hand, which is a settlement of some two and a half thousand Indians, only 280 children attend school.'

The fathers' efforts, as well as educating the Indians, served to instil in them a deep love for Colombia, which became apparent during the conflict with Peru, when a large number of natives enlisted to fight on behalf of the country. And yet the mission was criticized bitterly for not teaching the Indians about their Colombian identity.

Modern anthropologists who have not been above attacking the Catalonian Capuchins for this very reason would do well to recall the words of Dr Jaramilla, who wrote in 1932: 'It was a wise decision on the part of our government to put the education in the care of the Capuchin Mission, which has accomplished a difficult and important task. Their achievement can only be appreciated by someone who is familiar with all the circumstances; mere educational expertise would not be sufficient to understand the depth of knowledge required to penetrate the Indian psyche.'

Encouraged by the good results achieved in Colombia, the Capuchin fathers decided to establish more mission schools in the lower Caquetá. With the support of the administration of the Amazonas Commissariat, a very primitive school was opened in La Chorrera. Some forty Indians attended in the beginning what was to become the Capuchins' first boarding-school (*internado**).

Undoubtedly one of the principal reasons why Colombia had not been able to control the Caquetá was the total lack of roads and direct access to the interior. When the order of Catalan Capuchins agreed to take spiritual responsibility for the region, the then prefect, Fidel de Montclar, decided that it was time to establish means of communication between the Caquetá and other civilized nations. Only then could total and permanent

---

* The word 'boarding-school', translated in English dictionaries as *internado*, does not convey the correct impression. These missionary schools were opened to find a place to form and evangelize Indian children, who came there at certain times every year and had to be lodged and fed because the distances were too great between the settlements and there was nowhere else available. In the reserves, the Indians should now put up and run their own schools, provided there are sufficient indigene bilingual teachers.

conversion of the Indians to Christianity be achieved.

Fidel de Montclar conceived several projects, but the construction of a road from Pasto to Mocoa seemed by far the most urgent. Such a road would open up the Putumayo to the rest of Colombia. The distance between Pasto and Mocoa was eighty kilometres, and there was not even a path linking the two. The missionaries were too poor even to feed themselves sufficiently, so money was a great problem until the government eventually agreed to fund the project. Montclar personally supervised the construction, carried out by some sixty labourers, most of whom were Indians.

In this way the Capuchin missionaries gave fresh impetus to the establishment of a harmonious relationship between Church and state. It was a slow process and due largely to the fact that the missionaries were willing to undertake tasks which others refused even to contemplate.

Among the first missionaries in South America were Augustinians, called *Candelarias* in Colombia, who lived according to very strict rules. Many were also among the first martyrs, killed by Indians with hatchets and poisoned arrows. Members of many different religious orders worked in the interior at one time or another: monks of the order of St Francis of Assisi, Franciscan Capuchins, Missionaries of St Vincent de Paul, Franciscan nuns and, in the twentieth century, the sisters of Mother Laura from Medellin.

The Indians came to realize that the missionaries loved and cared for them, particularly for their children. At first they all walked around naked, taking and expecting gifts from the missionaries and offering nothing in return. But before long they began to lose their savage instincts and to acquire the habits of work and social intercourse. Trees were felled in the forest, and villages constructed; a textile industry grew from their new custom of wearing clothes, and they learned how to cultivate food. Their lives became relatively comfortable, and the missionaries built churches in which they could all worship Christ and his mother Mary. Frequently, however, missionaries and Indians alike were forced to move on from their settlements. Even the efforts of Simon Bolívar in the 1820s to restore lands to the native Indians were frustrated by Congress.

In 1842 a law was passed which provided for the establishment of mission colleges by Europeans. This enabled the Jesuits to come to Colombia in greater numbers, to found colleges in Bogotá and Popayán and to work in the Caquetá. Eventually Congress recognized the extraordinary strength and effectiveness of the Catholic missions and gave them adequate financial support. In 1912 a law granted an annual sum of 100,000 pesos in gold to be shared among all the missions active in Colombia.

One essential task was the construction of a road into the Putumayo. Only a few miles from Pasto, on the far side of the eastern Cordillera, lived thousands of Indians who for 400 years had resisted all the Church's attempts to civilize them. Several times missionaries had succeeded, by dint of heroic efforts, in clearing the forest, building villages and beginning to educate the natives. Each time unexpected events – wars, political changes – conspired to frustrate their efforts, forcing the missionaries to withdraw and destroying in a matter of days the results of years of labour. Once again the natives would return to their pagan beliefs.

Still, the missionaries continued to hope to build roads, to reach the Indians in the isolation in which they had lived for so many centuries. If they could only save the Indians' soul, they would turn them into useful members of Colombian society.

This message had finally penetrated to the government, although the help given towards the construction and maintenance of the mission schools was never enough.

Without the dedication of the missionaries nothing would have been achieved.

# 3   Hardenburg's Courage

While the missionaries were building roads and schools, word spread that in the rubber belt, especially in the Igaraparaná region, easy money could be earned.

The missionaries had achieved a great deal, but the lands of the Casa Arana, the Peruvian rubber company, were closed to them. The Peruvians did not respect their government's *modus vivendi* agreement with Colombia and Brazil, and closed their eyes to changes of which they disapproved.

In 1906 Colombia and Peru had ratified an earlier treaty entrusting the lands north of the Putumayo to Colombia and those on the south bank to Peru, but Julio Cesar Arana had his own ideas and persuaded Peru to give government protection to his possessions north of the river. The last people he wanted near his rubber trees were missionaries who might disrupt the work of his lucrative plantation.

The heart of the rubber-producing area was La Chorrera, the cradle of the Witoto tribe. It was said that the tribe had been decimated, due mainly to the harsh treatment it had suffered at the hands of the rubber-producers.

In that region, the first rubber-producer was a Colombian, Benjamin Larrañaga, who had come from Pasto with eight companions in 1897 on the Peruvian steamer *Cahuapana*. After studying the potential of the area, he settled in La Chorrera. In 1900 Julio Cesar Arana joined him, and they formed a company called Larrañaga, Arana & Cia. A year later, Larrañaga was dead, believed poisoned. His son and heir began to drink heavily, and he too died, in 1904. Arana established a new company in conjunction with Vega, the Colombian consul in Iquitos. Based in La Chorrera, the company, called Arana, Vega & Cia, took over the exploitation of the whole of the Igaraparaná.

Father Jacinto de Quito and Father Santiago de Tuqueres, who were on an apostolic expedition to the Caquetá, may have caused Julio Cesar Arana to fear that his cruel treatment of the Indian workers on the plantation might be discovered. Be that as it may, he decided to employ a geographer to advise him on the most suitable alternative areas for further rubber-production. The man he hired was an ethnologist by profession, a Frenchman from Poitiers called Eugène Robuchon, the son of a well-known French photographer.

In 1904 Robuchon signed a four-month contract to work for Arana which committed him to exploring all the rivers in the Putumayo, and all the adjacent territories from the Napo river to the Caquetá. He was to photograph the most noteworthy locations, the potential rubber plantations and the Indian tribesmen there. Robuchon came back to Iquitos with several albums of photographs.

Robuchon's life up to that time had been nothing if not eventful. He had found a young Indian girl near the River Inamban, fallen in love with her and taken her home to Poitiers, where he married her in a church wedding in 1902. After a year in France, the two of them left again, accompanied by Robuchon's dog Otello, for Mañaos in

Brazil, and from there penetrated as far as the River Igaraparaná, which was not recorded on any map at that time.

While he was in Iquitos, preparing for his third trip to survey the rubber plantations, Robuchon sent his wife, in the company of a nine-year-old Indian girl of the Boras tribe, back to his father in France for safety. He wrote that the poor child had been marked out by a sign on her breast for a cannibalistic feast when he had rescued her. Robuchon then returned alone to the Putumayo in 1906 and was never heard of again.

In 1906 rubber-production reached 540 tons, extracted by over 9,000 Indians. In 1907 Julio Arana travelled to London and established the Peruvian Amazon Co. Ltd, with a share capital of £1 million, of which he himself held £780,000.

Six years later a document, said to be written by Robuchon, was published in Barcelona, Spain, by the Peruvian consul, in which he praised Peruvians for their fair treatment of the Indians they employed, and accused Colombians of the very crimes against Indians hitherto attributed to the Peruvians. The document, written in Spanish, was also very critical of the Indians themselves. Given Robuchon's evident sympathy for the Amazon Indians, and his fears for his young wife's safety, it seemed likely that he had become too critical of the methods of the Casa Arana for his own good.

Meanwhile, back in Colombia, two young Americans, Hardenburg and Perkins, had left their work as engineers on the Cauca Railway to look for more lucrative employment in Amazonia. No doubt, the lure of adventure attracted them as much as the prospect of large earnings.

Walter Ernst Hardenburg's account of their trip, published in London in 1907 under the title *The Putumayo*, showed that, although they found great excitement in the primeval rain forests, they also witnessed unimaginable horrors. The wildest animals could not compare in savagery with the most ruthless of them all: man himself. A collection of articles and the book written by Hardenburg (all published eventually in London) later formed the basis of an official report, namely the *Blue Book* published in 1913 by the British consul Roger Casement, which led to the closure of the London company.

Walter Ernst Hardenburg was only twenty-two years old when he set about the task of collecting evidence against the rubber-traders. No doubt he spoke of his intentions when he reached La Chorrera; in any event, he was forced to leave and was taken to Iquitos on board a Peruvian ship. Once in Peru, he collected written evidence which he translated into English from Spanish and Portuguese. He obtained sworn testimonials from Peruvians and Colombians who had been employed by Arana and were willing to testify to the cruelties committed against the Indians. When these descriptions were published in London, accompanied by photographs, considerable public feeling was stirred up against Peru, to the extent that the British government requested the Pope to send a team of investigators to La Chorrera.

In 1908 a British explorer, Thomas Whiffen, arrived in Colombia to search for Eugène Robuchon, the French geographer last seen in February 1906 at the mouth of the River Cahuianari. He found traces which suggested that Robuchon had been there – a letter fastened to a tree, made illegible by the rain. But Whiffen found nothing conclusive. The Peruvians claimed that Robuchon had probably been eaten by Witoto tribesmen. The Colombian version of events was that Robuchon had indeed brought photographs to Iquitos, photographs showing half-naked Indians with heavy chains around their necks, just like those published by Hardenburg. The truth will probably never be known.

The following are extracts from Hardenburg's account, published in Roger Casement's *Blue Book,* bearing witness to even more horrific events:

The Barbadians (West Indians who worked at Casa Arana) were no savages. With few exceptions, they could read and write, some of them well. They were much more civilized than the majority of their supervisors; they were certainly more humane....

Every civilized employee receives, on arrival, an Indian woman to be his temporary wife.... From what I observed, their wishes would not be considered... the company agent would ask one of the many Indian women kept at the station if she wished to live with the new arrival. The man Dyall told me, in the presence of the chief agent of the Peruvian Amazon Company at La Chorrera, that he had served at some different rubber plantations and had different Indian women given to him as wives at various times. – These 'wives' had to be fed and clothed and, if there were children, they too had to be provided for. For this reason many of the Barbadians fell into debt....

The four main tribes were the *Witotoes,* the Boras, the Andokes and the Ocainoes, and there were other smaller tribes, such as the Ricivaros and the Muinanes.... The tribes' disagreements with one another never led to bloodshed; I believe the Amazon Indian is averse to shedding blood, and is thoughtless rather than cruel. Prisoners taken may have been eaten, for although the Indians do not seem to have killed to eat, as is the case with many primitive races, they did in the past eat those whom they had killed....

A half day's journey away from Caquetá, they caught an old Indian woman. Jiminez asked where the other Indians were. She lied and misdirected them.... He tied the old woman's hands behind her back. There were two trees standing about six feet apart. He made an Indian cut a branch to stretch between them. Then he hauled the old woman up so that her feet were dangling. 'Bring me some dry leaves,' he ordered one of the boys, and he put these under the old woman's feet as she hung there, then he lit them, and the woman started to burn. Huge blisters appeared on her skin and she began to scream. Jiminez had her taken down, but she was not dead; she was still screaming. 'If she can't walk, cut off her head,' he said.

The evidence of another Barbadian called Westerman Leavine confirms a statement given by Genaro Capero and published in the *Truth* in 1907. Three old Indians and their two daughters were murdered by Norman in cold blood and their bodies fed to the dogs. Leavine saw this take place. He described also the way Indians were left to starve to death in the '*cepo*', a wooden board with holes in it, like stocks, in which the feet were clamped. More than once he saw dead and putrefying bodies lying beside living prisoners. He remembered the incident recounted by Capero when an Indian Chief was burnt alive in front of his wife and his children. After that the wife was beheaded and the children dismembered, and all the bodies thrown on the fire. He also saw Norman cut an Indian woman to pieces with his own hands, after wrapping her up in a petrol-soaked Peruvian flag, setting fire to her and shooting her, because she refused to live with one of his employees. Capero had spoken of the ground being littered with skulls. Leavine said there were days in 1906 and 1907 when 'you could not eat your food on account of the bodies of dead Indians lying unburied about the place.' He frequently saw dogs eating them, dragging their limbs around.

He also confirmed another statement concerning a child being rammed head first down one of the holes dug for the house timbers.

Leavine said that more than once he had seen Señor Norman burn alive children who had refused to reveal the whereabouts of their parents.

These and other accounts taken from Hardenburg's book were officially confirmed in Sir Roger Casement's *Blue Book*. The Indian population had shrunk in ten years from 50,000 to about 10,000.

After the publication of the *Truth* and the investigation of the British government into its allegations, Roger Casement was knighted and, after a debate in Parliament in 1912, the company was dissolved.

The same year Pope Pius X denounced the atrocities committed at Casa Arana in an encyclical called *Lacrimabili stato*. Further investigations were carried out in the name of the Holy See, which confirmed Casement's accusations. This chapter of the story ends in 1913, one year before the First World War, with its increasing demand for rubber. Up until that time, no missionaries had been permitted to enter the land of the Casa Arana. However, acting upon the official request of the British government in 1912, the Holy See sent five Irish Franciscan fathers to minister to the spiritual needs of La Chorrera and neighbouring El Encanto.* The fathers stayed in La Chorrera throughout the First World War, while Casa Arana continued its cruel work. In 1918 they returned to San Antonio and reported that there were still over 6,000 Indians employed as slave labour in La Chorrera and in other rubber plantations.

The killings must have slowed down because of the missionaries' presence, but Casa Arana continued to make vast amounts of money at the expense of the Indian tribes.**

---

* The Holy See sent Father Léon Sambrok, their superior, Fathers Federico Furlong, Cipriano Byrne and Felix Ryan, and Brother Edwin O'Donell, accompanied by Father Alberto Gridilla, a Peruvian priest. In January 1914 Father Gridilla returned to Iquitos, and Father Furlong, who had become ill, was replaced by Father Leovigildo Olano.

** In 1918 the Irish Franciscan fathers reported that there were still some 6,200 Indians slaving in the rubber plantations of La Chorrera, Oriente, Entre-rios, Occidente, the branch of Emerayes, Atenas, and the branch of Charocamena (Witotoes), Sabana (Nonuyes and Muinanes), and Matanzas (Andokes and Boras).

# 4   The Caquetá

Under the terms of the 1906 treaty, the territories north of the Putumayo river belonged to Colombia and those to the south to Peru. However, after 1907 Peru's rubber plantations had encroached upon Colombian territory in the Caquetá. Even after 1918, after news of the scandal had reached London, the Casa Arana continued its work in the rubber belt, destroying nature and natives alike.

Eventually the Colombian Senate made a declaration:

1) that the governments of Great Britain and the United States of America had addressed themselves to Peru, demanding that Peru should cease the atrocities committed by Peruvian citizens against the Indians of the Putumayo, and that it should punish the perpetrators of those crimes against humanity and against civilization, in the mistaken belief that it was Peru which was sovereign of the regions in which such crimes had been committed;

2) that the so-called *Blue Book*, by Sir Roger Casement, although published by the government of Great Britain, was based on the same error, and

3) that the accusations of the world's press against the government of Peru implied the same erroneous belief.

Further, the Senate decided:

1) once again to confirm the legitimacy, in fact and as of right, of the titles granting Colombia sole sovereignty of the Caquetá, of the shores of its rivers and of their tributaries;

2) to declare that from 1901 to the present date, first rubber-producers from Peru and then authorities of that country had proceeded to violate and settle in Colombian farmlands. This was in disregard of the rights of Colombia, which had not renounced its dominion over those territories.

3) that if the Colombian government had not been kept away from those places, acts of violence would certainly not have been committed; furthermore, Colombia would have sheltered and protected its citizens, bringing Christian civilization to the affected tribes, as it intended now to do as soon as possible;

4) that it (the government) protested against the usurpation which had temporarily deprived Colombia of its legal ownership of these regions;

5) that if, in order to recover or defend these regions, conflicts should occur with invading Peruvians, the fault would not be with the Colombian nation, which had justice on its side, but entirely with the invading forces.

As most of the rubber trees in the Putumayo had been destroyed and the Witoto Indians decimated, the Peruvians of the Casa Arana came to the Caquetá, no doubt to re-enact the same tragedy. The Colombian government began to see the need to establish a military presence in the south of the Caquetá and therefore sent General Isaias Gamboa, who had long been campaigning for this, to command a small garrison at Pedrera, below the port of Córdoba. The general maintained that colonization of the Colombian part of Amazonia would establish once and for all Colombia's sovereignty there, and prevent the Peruvians from penetrating any further. His other purpose was to ensure that Colombian ships could navigate the Caquetá river, particularly the section between Mañaos and Araracuara, which could be threatened with an attack by Peruvian forces.

The general, profoundly convinced of his cause, requested 500 soldiers, a battleship and sufficient funds to provide rations for his troops. Despite his pleas, support on this scale was not forthcoming. The expedition set off from Barranquilla with a hundred men, scanty provisions and no financial support. General Gamboa was promised further help when he reached Martinica, where a Colombian cruiser, *Cartagena*, was under repair. He and his soldier marched from Barranquilla to Puerto Colombia and from there on to Belen de Para. From Belen they went on to Mañao and then to Tefé. But in Tefé the expedition was halted for thirty-four days, because there were no boats to carry them to La Pedrera.

In Tefé General Gamboa met a representative of the Brazilian government from the frontier town of Yaperé, who gave him secret information about orders received from the Peruvian government representation, in Rio de Janeiro, to permit the free passage of all Peruvian boats and armed forces. General Gamboa notified the Colombian War Ministry and begged them to send enough men, arms and provisions for him to deal with the situation. He embarked for La Pedrera on 31 March 1911. The General decided to stay in La Pedrera, where the Colombian frontier post was situated. It was also the nearest point to the border and in the best strategic position for defence.

Once again the General asked for support from Bogotá. He notified the authorities of the poor health of his troops, and of the defence work they had already carried out, and he told them about the shortage of food. He had heard that military forces had left Lima with orders to invade and occupy the north-western shores of the lower Caquetá. He asked urgently for 200-300 experienced men to defend the Caquetá; its invasion would have serious consequences for the integrity of Colombia.

On 10, 11 and 12 April 1911 the Battle of La Pedrera took place, between the remaining sixty-five men under General Gambao, helped by some twenty men from the military post of the port of Córdoba, and 450 Peruvians with four armoured launches under Colonel Oscar R. Benavides, who was later to become President of Peru. The battle continued until four Peruvian launches penetrated the rapids of La Pedrera, disembarked and occupied the entire village.

While General Gambao and the then remaining thirty-five Colombian soldiers took refuge at the Brazilian frontier post of Villa Betancourt, the Peruvians carried the Colombian General Valencia and twenty-three soldiers as prisoners to Lima, Peru.

A new treaty, negotiated by the respective consuls of Colombia and Peru in Manaos, ended the conflict. In July 1911 it was agreed that the Peruvian forces should leave the area of the battle. In this the status quo was ratified, and Article 50 declared that, 'The presence of Peruvian authorities and Peruvian forces in an area does not signify the recognition of Peruvian rights over that area.' Shortly afterwards, the Peruvian army withdrew from La Pedrera, leaving the Casa Arana to continue its crimes against the Indians without fear of punishment.

As mentioned before, one of the earliest and greatest problems in carrying out building work in the Amazon region is the fact that there are no natural stones, no quarries, in the entire Amazon valley. However, the Caquetá and the Putumayo had, and have, stones enough. 'La Pedrera' (*pedra*, stone) means 'stone' or 'quarry', and for centuries it provided stone axes for the warring tribes. The Indians who lived at and near La Pedrera were mainly the Andokes. In 1820 the German Martius announced the existence of petroglyphs (prehistoric stone carvings) in the waterfalls of Cupati, Araracuara. The Indians believed firmly that the region had been formerly inhabited.

A French explorer, a missionary father by the name of Tastevin, discovered the carvings first in 1923 in La Pedrera. He reported that the inhabitants of Caicera and Tefé obtained their provisions of stones for liming in La Pedrera. With the coming of the republic, the territories of the Caquetá and Putumayo were joined to the state of Cauca.

Meanwhile, in 1922, another treaty was signed between Colombia and Peru, in an attempt to resolve the territorial disputes once and for all. This, the Salomón-Lozano Treaty, finally established the frontiers between the two countries.

Herein Colombia declared that all territories between the western shore of the Putumayo river, towards the east of the River Cuhimbe, and the frontier between Colombia and Ecuador in the Putumayo and Napo valleys, belonged to Peru. Both Colombia and Peru acknowledged in permanence the rights of territorial transit and free navigation on all their rivers and tributaries. Finally, Colombia appealed to the good offices of the USA as mediator, and the three nations signed a treaty in Washington in 1925, according to which Colombia and Peru ratified their own treaty and recognized the official borderline of Tabatinga-Apaporis between Colombia and Brazil. Brazil allowed Colombia the right of free navigation in perpetuity on the Amazon and other rivers common to both countries.

The government of Dr Abadia Méndez proceeded to occupy 'the Amazon Trapeze'. The additional land now belonging to Colombia, its southernmost point the Amazon river, has indeed the shape of a trapeze. It is bordered on the eastern side by Brazil and on the western by Peru (The treaty of limits gave to Colombia some additional 20–25 square kilometres calling it now 'trapeze'.) An army was brought to the area, and the process of colonization began. By a government resolution of April 1930, the sum of 5,269,000 Colombian pesos was granted towards the maintenance of a Colombian garrison in the Amazon. In August 1930 the Colombian Army took full possession of the Amazon Trapeze.

The little town of Leticia already existed at the point of the Trapeze. The Peruvian engineer Manuel Charcón had begun some construction in 1867 and had named the town Leticia in honour of his fiancée, Leticia Smith. Notwithstanding, the fickle lady married the British consul in Iquitos, Peru.

In 1907 it had been decreed that all steamships travelling to Iquitos should have a medical practitioner on board, so now a medical school was built in Leticia, which considerably enhanced its population and reputation. The same year a Catalan priest, Father Pedro Prat, came to live in Leticia. Later he moved to Nazaret, today Porto Amalia, on the Yavari river.

In subsequent years, various treaties between Brazil and Colombia formalized the frontier between the two countries in this area, culminating in the Act of Iquitos of 1930.

In that year the steamer SS *Narino* arrived in Leticia. On board were Colonel Acevedo; the first commissioner, Abdon Villareal; Father Bartolomé de Igualada, the military chaplain; Gabriela Marin, Leticia's first teacher, and thirty-two civilian farmers.

Father Bartolomé described the impression on arriving in Leticia: 'There is a wireless mast which receives signals broadcast from all over the world, and there are some twenty houses where Peruvians live. One can also see the foundations of a large fort with three canons, all that remains of the building destroyed by the Peruvians about fifteen years ago, as they could use the bricks for construction work....'

On 17 August 1930, having officially taken possession of the land from the Peruvian authorities, Colonel Acevedo gave the order to hoist the Colombian flag for the first time. The duty post was established and five prefabricated houses, brought by the steamer *Neiva*, were put up. In 1931 the Amazon Administration Office was formed, and the former Colombian consul in Iquitos, General Amadeo Rodriguez, was named military and civil commander.

Feeling that all was in order, in 1932 the garrison of Leticia, composed of thirty-five men and one officer, was transferred to El Encanto. Father Bartolomé de Igualada, one of the first and most important Capuchin missionaries from Catalonia, was given spiritual responsibility for the entire Amazon region, while residing in Leticia. So all the necessary arrangements had been made and the rule of order prevailed when, on 15 July 1932, an event occurred which shocked the nation. The assault and occupation of Leticia by Peruvian forces became justly known as 'the Scandal of Leticia'.

Twenty Peruvian soldiers and 200 Peruvian civilians, under the command of Oscar Ordoñez, invaded the town. The Colombian naval commander Villamil and many of the civilian population were transported to Benjamin Constant in Brazil. But this time Colombia defended itself on all fronts, as the treaties had established Colombian rights and frontiers.

On 25 June 1933 a League of Nations Commission was established in Leticia at Colombia's request, to take over the administration of the whole area.

At that time the population of Leticia consisted of some seventy Colombian workers and eighty Brazilians with their families, making a total of 402 inhabitants, living in just forty-one houses.

Another Catalan, Father Lucas de Batet, took charge of the parish of Leticia. On 24 December 1933 he performed the first baptism in Leticia.

General Ignacio Moreno Espinosa was nominated as the new commander (*intendente*) of the Amazonas. On 24 May 1934 the Salomón-Lozano Treaty of friendship and co-operation between Peru and Colombia was ratified, and regular navigation could again begin on the Putumayo. The enterprise Navesur with its three steamships, reopened navigation on the Putumayo. On 19 June 1934 the League of Nations Commission in Leticia transferred its sovereignty over the entire territory back to Colombia.

General Carlos B. Muñoz was named mayor of Leticia. Father Lucas de Batet blessed the first stone of Leticia's new church, the parochial temple of Nuestra Señora de Paz (Our Lady of Peace) in July 1936.

The Brazilian consulate was opened and in late 1938 an agreement of co-operation concerning duty questions between Colombia and Peru was signed. Peru also opened a consulate in Leticia.

In 1935 Father Bartolomé de Igualada had brought from Manaos, Brazil, to Leticia prefabricated buildings of metallic construction, so that there were now more houses available. At that time the census of the entire intendency of Leticia recorded 6,414 persons, of whom 2,232 were Indians.

While Brazil began to develop its Amazonian lands by building an impractical Transamazonian Highway, Colombia chose a different method of colonizing its vast territories. It asked repeatedly for spiritual aid from Catholic missionaries, hoping they could bring education to the native population, no easy undertaking because these missionaries were later on accused of not respecting the original culture of the different Indian tribes in the Amazon region.

The important question of *how* the natives reacted upon the permanent political disputes between Colombia and Peru and what happened to the different tribes who lived in Amazonia, of whom many were completely extermi-nated, seems to have been the basis of an investigation of the tribe of Mirañas during the years 1971–6.* Its author

* Mireille Guyot in *Journal de la Société des Américanistes*, volume LXVI.

feels that these Indians in their myths and legends assimilated intellectually the historical happenings which affected their very existence. Guyot is referring to the introduction of trade and commerce by the white nations into the closed world of the Indians, up to then completely unknown in Amazon Indian history.

The Mirañas occupied two small hamlets, separated by about 400 kilometres; the only connection between them was the course of the Caquetá river until the coming of airplanes.

The houses of the Mirañas were dispersed over some eighty kilometres, beginning at a place called Mari-Manteca, at the right shore of the Caquetá river, until the last houses at a distance of more than 200 kilometres from Araracuara.

Although the Yukuna and the Matapi at the north of the Caquetá river, linguistically quite separated, offered to the Miraña (who had fled before the rubber-dealers of the Casa Arana) the occasion of new land purchases, the great majority of the Miraña returned to their own land. They call themselves *'gentes* (people) of the centre of the world' and 'people of God', a grouping which comprises the tribes of the right side of the Caquetá river, namely the Witotoes, Andokes, Boras and Muinanis, in opposition to the *gentes* of the left-hand shore, whom they call 'people of the animals', because they sacrifice animals to their spirits of protection, whereas the Mirañas and their neighbours of the south offer only plants and vegetables to their divinities. The Mirañas call all white people 'labourers of God', because they 'made it all'.

Nowadays the Mirañas wear shirts and pants and have exchanged their blow-pipes for guns. The women dedicate their early morning hours to their *chacras* (farming work), while the men go hunting or fishing or manufacture certain objects. At about 9 p.m. the women must retire with their children, while the men sit in a small, closed circle, upon very low seats or on their heels. Then the men begin their important 'recreation of the world', together with the forces of beyond, called by them '*más allá*'.

In a myth it is said that, '... at first, at the mouth of this river [Caquetá], near Tefé, was the maggot of yucca [a grub which devours the leaves of the yucca], which seemed to be a divine grub ... then came a large snake of the trading of the *dantas* ... there also was a spirit making a noise just like a husky, so hoarse, the master of a river below La Pedrera, which was a spirit of evil and of the business of war ... and then there was that spirit of the dead souls, terror of the bottom and seat of the water, of La Pedrera, which evil spirit had its house in the waterfall, a house of evil and great terror....'

After telling two myths Guyot states: 'This myth of the Caquetá, which unrolls like a litany and seems to be the only one recalled in this form by the Mirañas, begins at the mouth of the Caquetá river at Tefé, Brazil, and rolls upstream until reaching Araracuara in Colombia, whence is the limit of the territory of the Mirañas. As a whole, the legends are a testimonial of the reactions of the Indians upon their introduction to having to trade with white people.'

It seems quite remarkable that – after all these tribes had suffered when exploited by the Casa Arana – their legends express no bitterness, but only admiration of the white men, 'because they made it all'.

The first detailed description of the tribe of Miraña we owe to the German traveller Martius (1820). He distinguished two groups: the Miraña-carapana-tapuio living near Araracuara and on the shores of the River Caquetá, and the Miraña-uiranasu-tapuio (Koch Grünberg) on the shores of the River Cahuinari. The two languages are not related. The first group belonged to the linguistic family of the Huitotos (Witotoes), the second to the Bora-Mirañas.

In 1820 Martius reported that there were some 6,000 Mirañas living between the Yapuri and the Içá. Puerto (Port) Miraña was well known by the Hamelucks and the Lusobrasilians for their trade with slaves. The captain, who was wearing a shirt and pants but ignored both the Tupi and the Portuguese languages, bought Indians and fought with other Indians to obtain slaves. His authority was based only upon this trade.

From the information obtained from Bates (1850) and Marcoy (1873) and others, the traffic of Miraña slaves lasted during the entire nineteenth-century: 'From that trading between civilization and barbarism resulted a number of young Miraña slaves of both sexes, which one may see in the villages and cities of Amazona from Alvaraes until Barra de Río Negro' (Marcoy, 1873, p. 451).

# 5  Dr Paul Rivet

It is often forgotten by Europeans that Latin America was colonized by their compatriots and industrialized by their kinsmen. Spaniards, Portuguese, the British, Germans and French care little for not-so-distant links. They prefer to forge newer, more fashionable connections with impoverished African lands, former colonies whose music, art and rhythm fire the imagination in ways Latin America has failed to do.

An attempt to halt this trend was made in Paris at the beginning of this century by Dr Paul Rivet, when he created the Société des Américanistes, the Latin-American Society, at a time when French, and not English, was the second language for all young, well-educated South Americans. France was then also a theatre for great events in the history of the world: the abolition of slavery was first announced in the National Assembly in Paris, by a Negro and a mulatto. It was to be heard throughout the free world, although the news took a long time to reach American Indian ears. In South America it was the Indian who paid dearly for the wealth accumulated by white men in the rubber plantations. Perhaps the Indians had been happy, living in their primitive way close to nature. All the land in South America, and, indeed, in North America, had belonged to the Indians until white men took their land and often corrupted their way of life.

The origin of Indian tribes in South America had not been satisfactorily established. Paul Rivet's theory was that they came from a number of different sources: from Asia, over the Behring Sea in the Antarctic and from Polynesia and Melanesia. He did not believe there were any autochthon (aboriginal) Indians in South America.

Paul Rivet was born in Wessigny in France, in 1876, two years after Laura Montayu, who founded a congregation in Medellín with the 'constitutional purpose' of saving the souls of the Indians. He studied surgery and military medicine at the national school of military medicine in Lyons and graduated at the early age of twenty-one. Soon after his graduation he was chosen by the French government to participate in a scientific mission to South America to take a fresh measurement of the meridian in Ecuador, on the basis of Condamine's eighteenth-century calculations. During his six years in Ecuador, he practised medicine, as well as carrying out his scientific work. He came to know the people of Ecuador well and developed a profound affection for South Americans, especially for the Andean population.

Dr Rivet resented the fact that the Indians were regarded as savages; from his personal experience, having lived with groups of Indians in Ecuador at various times, he had often found them to be people of great intelligence. (No doubt he was also influenced by his admiration for Jean-Jacques Rousseau's 'noble savage'.) He began methodically to collect scientific information about Indians and came to the conclusion that there were indeed no significant differences in mental capacity between white men and Indians.

Based on his anthropological studies, Rivet dedicated himself to encouraging greater tolerance and understanding among peoples. It was in Quito that his life-long views on anti-racism and socialism were formed. From Quito he submitted his first report to Paris in 1903, a study of 'Indians living in the Rio-Bamba area' of Ecuador. At that time France was achieving great scientific development, guided by Buffon, Lamark and Cuvier. Rivet began to collect data and when, in 1907, he returned to France, he took with him not only the results of his scientific experiments but also a large archaeological collection. He was accompanied by his wife, Mercedes Andrade Chiriboga, a high-society lady from Quito, who was to be his lifelong companion. Back in Paris, Dr Rivet abandoned the medical profession and began to work with the Musée Nationale d'Histoire Naturelle to classify and study the large anthropological, ethnographical, archaeological and linguistic collections which he had brought from South America. Twenty years later he was nominated to the Chair of Anthropology.

At the same time nominated Director of the Museum of Ethnography at the Trocadero in Paris, Dr Rivet began to renovate what had become a very out-of-date museum, converting it into an ethnographical institution which he named 'the Museum of Mankind' (Musée de l'Homme). This was to become also the seat of the *'Américanistes'*, their journal and publications.

During the First World War Paul Rivet interrupted his life of research and served again as a physician in the French Army, working as director of the Military Hospital. His wife was wounded driving an ambulance at the battlefront. He continues to publish the results of his work during fifty years of his life, interrupted only by the next world war.

The Musée de l'Homme, as it still is called, was finally opened in Paris in 1935. Dr Rivet gave a series of lectures (the Chair of Anthropology which he held had been attached to the former Musée d'Ethnographie), during which he recommended all young *Américanistes*, as they were then called, to study the languages of the American Indians.

This was no easy task; with the notable exception of the Mayan hieroglyphs, most of the languages were spoken only. However, Paul Rivet was profoundly convinced of the value of South American pre-Colombian culture, and also determined to transmit this enthusiasm to his pupils, so they would continue his work.

Early in his Parisian career, Rivet must have learned of the evil doings of the Peruvians in the Putumayo. With his South American interests, he was in regular contact with the Colombian Embassy in Paris, so was aware of the mission of five Irish Franciscans appointed by the Pope, at the demand of Britain, to investigate and put a stop to the cruelties committed against the Indians in the rubber plantations.

Meanwhile Rivet's academic work continued. He published his most important work, *Origins of the American Man*, which went into four editions and greatly influenced the *Américanistes'* theories.

Shortly before the Second World War his belief in the freedom and equality of all men was to take a different and – for his own safety – a very dangerous form. He was interested in political thought and was friendly with Léon Blum and Guy Môquet, among others. Under cover of his scientific activities, he was president of a society of anti-fascist intellectuals, which brought together some of the most courageous people in France. Before and during the war, he organized 'The Conspiracy of the Museum of Mankind' against the occupying German forces.

All the other members of this Conspiracy were killed by the Gestapo, but Dr Rivet, thanks to his friendship with Eduardo Santo, then President of Colombia and helped by the Colombian Embassy in Paris, was able to escape at the last moment.

He spent the greater part of the war in Bogotá,

continuing his efforts to bring Europe and South America closer together, and to establish the study of Indigene Anthropology as a serious academic subject. He helped to found the Anthropological Society in Bogotá, assisting young Colombians to gain a new insight into the craftsmanship and skills of early Indian tribes.

These same Colombians were later to join with foreign anthropologists to salvage many of the beautiful gold artefacts made by the Indians, in particular ornate funerary ornaments stolen from ancient graves, which were in great danger of being melted down for bullion. Dr Rivet inspired young Colombians to see their great artistic worth, where the Spanish *conquistadores* had been blind to all but monetary value.

The conversion of an old prison building in the heart of Bogotá into a museum of ethnology, archaeology and anthropology was also inspired by Paul Rivet. He was assisted in this, and in the management of the National Ethnological Institute, by one of his protégés, Dr Luis Duque Gómez. Thus Rivet and his Colombian friends succeeded in an even more significant task: they convinced the Colombian State Bank (the Banco de la República) to fund the now world-famous Gold Museum (Museo del Oro) in Bogotá, the splendour of whose exhibits remains unequalled.

During his absence from France in the war years, Rivet continued to support de Gaulle's government in exile, raising funds in Colombia which he sent to London. But, impatient of spirit, he left Bogotá in 1943 for Mexico, where he was cultural attaché in the Free French government, holding the rank of consul for the whole of Latin America.

After the end of the war, he returned to Paris to manage the Musée de l'Homme. By 1946 he was the head of the French delegation of UNESCO, in Paris, and was elected its president in 1953.

Rivet died in 1958, after a lifetime dedicated to the investigation of ways to improve the conditions and standard of living of the Latin American Indian.

# 6  Nofuico – La Chorrera

One of Paul Rivet's most brilliant students in Bogotá was a Capuchin father from Castelldelfels in Catalonia, Father Marcelino Canyes. Already in 1933 Father Marcelino had founded in Bogotá the *Centro de Investigaciones Linguisticas y Etnograficas de la Amazonia Colombiana* (the Centre for Indigene Linguistic Studies, CILEAC) for research on ancient indigenous languages and offering a library with literature about Indians, general information, descriptions about the regions where they lived. Like Paul Rivet he was convinced of the need for missionaries to speak, or at least to understand, some of the languages of the Indian tribes in Amazonia. He realized that to know all these different languages was quite impossible, so he taught the Indians in Spanish, believing that only after having received primary education could they become convinced Christians, Colombians, full citizens, capable of earning their living amid white people. Father Marcelino recalled that in Mallorca the tertiary Franciscan missionary Ramón Lull had been the first to open an institute for the study of languages. Also Ramón Lull had himself studied Arabic so that he could attempt to convert to Christianity the Arabs still remaining on Mallorca.

After publication of Hardenburg's articles and book on the Peruvian Casa Arana, the Amazon Rubber Company in London had been forced to close. So the production of rubber, which was still in demand, continued on a smaller scale, interrupted only by occasional frenzied attacks on the rubber trees by enslaved Indians whose misery continued.

*Truth* magazine and the Anti-Slavery Society had published Hardenburg's accusations with great hesitation, because of the issues and the large amounts of money involved. However, Parliament had acted only upon the publication of Sir Roger Casement's report. Already, in 1876, seeds stolen from Hevea trees in Brazil had been planted in Ceylon and then brought to Indonesia (especially Sumatra) to begin British plantations over there. During the First World War artificial latex was produced in Germany but only in 1939 was the invention fully developed, while in Latin America natural rubber-production dropped more and more.

As for Colombia, before 1933 there existed in El Encanto a garrison of some seventeen soldiers with about thirteen hectares of land soon to be cleared. Then, on 22 January 1931, a major conflict with Peru broke out, called 'the Battle of the Versings of the Caraparana' during which Candido el Leguizamo died of his wounds. The site was named Puerto Leguizamo.

Many of the Indians, especially those belonging to the Witoto tribes who had been deported to work on the rubber plantations in Peru, returned to Colombia after the end of the war with Peru. The Capuchin missionaries knew what it was to be torn away from a beloved country; they were not slow to take up the challenge of rehabilitating these Indians.

Meanwhile, in Europe, the Second World War was soon to break out. The Roman Catholic Church in Spain was under increasing attack by the Republicans. In Madrid and

Barcelona priests and nuns were shot in public. Nobody intervened in their favour, nor did anyone prevent the damaging of Barcelona's beautiful Gothic cathedral. Within the hearts of some of the fervent Catalan Catholics something had broken – the very roots of their love for Spain had been destroyed. And so some took refuge in Mallorca, less affected by the frenzy of German Nazism than was the mainland. From there, some of the Catalan fathers and Capuchin friars followed an old missionary tradition and asked to be transferred to Latin America. They had heard that there was a great need for them in the valleys of the Amazon, the Putumayo and the Caquetá, watering the Colombian part of Amazonia. They would heal wounds and bring Christianity to the Indians as missionaries.

As to La Chorrera, or *Nofuico* as it is called on the language of the Witotoes, claiming here the origin of their tribe, it can be seen in the beautiful church of the Chorrera *internado* that the main altar rests quite solidly on two cut trunks of a Hevea rubber tree, showing the side scars cut into them for the extraction of rubber fluid. The Capuchin Father Cristobal, much later director of the *internado* for some twenty years, wrote that the tree trunks served as a reminder of past sufferings of the Indians. 'Christianity does not destroy what it encounters,' he wrote, 'but dignifies and elevates all it touches.'

The Peruvians had named the place La Chorrera, because of the *Chorro* (jet of water) which divides the River Igaranaparaná in two, falling from a height of about ten metres and cascading along for about a hundred metres, before cutting through the stone, thus diverting the normal course of the river.

After the last battles of the Putumayo, at the beginning of the year 1933, General Amadeo Rodriguez took possession in the name of the Colombian nation of all the installations of the Casa Arana. The Capuchin Father Estanislao de las Corts arrived at La Chorrera on 2

November 1933. On 22 November General Moreno handed over to Father Estanislao all the remaining buildings of the Casa Arana, 'so as to instal in them an orphanage and a hospital'.

Father Estanislao de las Corts, who was originally from Barcelona, had worked in Pasto and in Sibundoy before becoming the head of the mission in La Chorrera. He died in Bogotá in 1940, but his work in Colombia has continued in the mission to this day, and his name, highly honoured, is commemorated in an important part of the Spanish city of Barcelona. There is also a town in Catalonia, Spain, named Igualada. The first great Capuchin missionaries have not been forgotten in their home country.

At first the missionaries found the Indians very difficult to communicate with: those who had fled from the Peruvian rubber-dealers were really terrified of white men, and it took a long time to persuade them to return to La Chorrera. Most of them did eventually come back, reassured by the friendliness and peaceable nature of the missionaries. Although the Capuchin fathers could offer them little earthly nourishment, the hungry Indians soon learned to put their trust in the fathers' kindness.

The true story of La Chorrera is little known or forgotten, for it does not reflect well on any of the governments involved, whether Peruvian, Colombian, British or even that of the Vatican. There was (up to 1988) no attempt to give financial compensation for the awful crimes committed against the Indians, the rain forests and the wildlife of the Amazon.

The first Capuchin mission-foundation of Amazonia, La Chorrera, was opened in 1933 with very modest means. The official decree was that: 'A certain charity mission will be established in La Chorrera, to be called the Orphanage of La Chorrera. There will be two sections: an orphanage for the care of Indian children of both sexes, and a hospital for the care of Indians of all ages.'

Father Estanislao de las Corts had been nominated by his order to accept the donation of the old Casa Arana buildings and the surrounding land in the name of the Capuchin Province of Catalonia. In December, 1933 he was appointed the first director of the school, with Doña Gabriela Marín as a teacher. Many children in the area died of a smallpox epidemic; those who survivied – sixteen boys and six girls – were brought to the mission school to be educated.

The first problem Father Estanislao faced was shortage of food. There were no hens or pigs left, and very little yucca, the plant which formed the Indians' staple diet. There was only a little flour, some dry biscuits and some coffee. The children had been used to chewing the leaves of coca plants, or anything else they could find, to dispel their hunger; the discovery of any meat was a source of general rejoicing.

The children lived and slept in an abandoned forge. The missionaries had to provide everything: the children had no clothes, no crockery or cutlery and thought of nothing but their empty stomachs. Despite these difficulties, Father Estanislao began to hold classes. Eventually he succeeded in teaching the children some of the catechism.

After a while the missionaries appealed for help to the government in Leticia. They had already tried Bogotá in vain. The reply from Leticia wished the mission well in elaborately pious terms but regretted that no funds were available 'in the current financial year'. The Capuchin fathers were on their own.

Meanwhile the indigene boys and girls slept together in the same dormitory and learned to eat and wash themselves in the white man's way. Occasionally the bones of a *danta*, a tapir, would be brought in, and then there would be soup to eat.

More and more children were brought by their parents to the mission school. Many came from Tarapacá, having fled during a great conflict there. These Indians had not been converted; indeed, they were totally ignorant of Christian ways.

It was a wonderful day for Father Estanislao and for the mission when, on 7 November 1934, the first baptism took place in La Chorrera, and Teresita Kassi, a 40-year-old Indian woman, was received into the Catholic Church.

In July 1935 Father Bartolomé de Igualada carried out an apostolic expedition to the Boras of the Cahuinari river. In August, he met another Capuchin, Father Javier de Barcelona who was on his way to find the very timid Mirañas of the Pamá river. And then, one afternoon, Father Javier de Barcelona arrived from El Encanto on a mule. He did not stay long before realizing that there were neither hens nor pigs nor yucca to eat, so he left. When he returned, with a fellow Capuchin, Father Placido, they brought with them pigs, goats, donkeys and cattle.

Further help came when, in October 1935, the first 'Lauritas', the missionary sisters from Mother Laura's congregation in Medellín, arrived at the school. Meanwhile the Indian population, scattered by the terrors of the Casa Arana, began to resettle in the region, and the missionary work continued with a fresh impetus.

In February 1936 Father Plácido de Calella, the regular superior of the mission, visited La Chorrera and appointed Father Javier its director, to replace Father Estanislao, who was unwell.

Some years later Father Javier wrote of his experiences in trying to obtain food for the school:

I had ordered seeds from a variety of sources, and I was astonished to learn that those ordered from the Government's administrative offices had simply been thrown away; they were never sent to me, nor to anywhere else in this Ministry. Later, I had to go to Bogotá for an operation. I brought back seeds, medicines and also a number of cats, as my personal luggage allowance. In

Leticia we managed to obtain some goats; we bought some sows in Tarapacá, hens in El Encanto (and very expensive hens they were), and cattle in Puerto Asís and in Caucayá. Pigeons from Manaos, hens and rabbits from Bogotá, more seeds from everywhere. However, half of this motley baggage was lost on the way back to La Chorrera.

I have no explanation for the losses, and I do not know why I was forced to do everything myself, but the result of all these efforts is that every single Indian in La Chorrera now has hens, many have pigs or goats, and together they are growing enough food to make the orphanage all but self-sufficient. If only there were adequate roads, their produce could even be sold.

More and more Capuchin missionaries came to the Amazon. In 1931 a young priest arrived in Sibundoy from Catalonia. He was Marceliano Canyes, Father Marcelino's younger brother. Father Marceliano stayed in Pasto and in Sibundoy twelve years in all. At first he was the parish priest and then the Vicariat's administrator, especially responsible for all the territories leased by the Capuchins to the Indians until 1946, when they were given to the Indians. From 1948 to 1952 Father Marceliano was deputy inspector of education in the Caquetá, based in Florencia, as well as continuing to carry out his duties as parish priest.

Father Estanislao had completed the road from Pasto to Mocoa before leaving the region, an old and tired man. His successor at La Chorrera, Father Javier de Barcelona, was kindness personified: he radiated love for the Indian children, and his work for them bore fruit.

As they still had no assistance from the government, the missionaries resigned themselves to the fact that they would have to find their own solutions to their problems. The only way to grow enough food was to teach the Indians to use more modern methods of cultivating the land and rearing livestock.

A new mission school was built. Part of its land was given to the village for the construction of an administrative centre, an electricity generator, a hospital and a telecommunications centre which would link La Chorrera with the whole of Colombia. Irrigation canals were dug, and new houses were planned. (The land, of course, really belonged to the Indians; the buildings had been given to the Capuchins by the Colombian Army under General Moreno, in an official act.) The Indians were capable carpenters, using skills learned long ago from their evil Peruvian masters at the Casa Arana.

In 1947 the mission school at La Chorrera had 146 pupils: sixty-seven boys and seventy-nine girls, now with a daily food budget of one fifth of a Colombian peso for each pupil.

In 1949 Father Javier de Barcelona consecrated the new church at La Chorrera. The following year he visited the source of the River Igaranaparaná, where the Indians believe that their ancestors are sleeping and do not wish to be disturbed.

At that time, all transport was water-borne. Not until 1970 did the Colombian government decide to build a landing-strip, so that La Chorrera could be reached within a few hours, rather than days, from Leticia. (Note that this decision to build was not translated into action until 1985, and at a cost of some 40 million Colombian pesos.)

# 7   The Witotoes

The name 'Witoto' is a corruption of 'Huitoto' ('Uitoto') first used by the German anthropologist Professor K.T. Preuss:* 'Witotoe' comprises a multitude of communities. The scientific name is 'Murui-Muinane', while 'Witoto' or ('Huitoto') signifies for some 'slave' and for others a meat-eating ant.

The so-called Witotoes are the second largest group. Elizabeth Reichel Dussan in her recent publications speaks of about 70,000 Indians actually living in Colombian Amazonia. In 1980 there were about 2,800 members of the Witoto living in Amazonia. The Murui-Muinane claim to have originated in La Chorrera, at the source of the River Igaraparaná, a tributary of the Putumayo. They are known to have their own distinct culture, although there are some similarities with the Ocainas, the Andokes and the Nonuyas. Unlike other tribes, they use nets for hunting big game. When finally an animal is caught, they beat it to death. They do not smoke pipes but mix cooked tobacco with water, then roll the mixture (called *ambil*) up in the

---

* Prof. K. Th. Preuss, Göttingen, A.F., West Germany, *Religion und Mythologie der Hitoten 1921–1923*.

form of a cigar. They are said to have practised cannibalism in former times, but there is no certain proof of this. The dialect of the tribe is Bue.

Their lifestyle has been documented by ethnologists. It has thus been noted that they construct huts, called *malocas*, for several families to live together. The shape of these *malocas* is unusual in that they are oval and have high, enclosed roofs. According to the Swiss anthropologist Jürg Gasché, there are female and male *malocas*. This duality is characteristic of the Murui-Muinane. Their history and culture were also expressed in their statuary, carved in wood and depicting tribal ritual. Unfortunately the tradition of carving statues has now disappeared, for many reasons, from the Murui-Muinane tribe.

Certain symbols have the capacity to absorb within them the totality of a culture. These are dominating symbols, while others are considered secondary symbols, when they express only certain aspects. With the Murui-Muinane, the yucca (*Manihot esculenta*) has the capacity to absorb the entire cultural universe. Basically, the Murui-Muinane man considers himself as yucca, and from this identity he constructs his symbolic universe, deriving his material existence from the food, yucca. The maximum expression of his social identity as a Murui-Muinane yucca is offered in the man known as a buinaima. It is as if the yucca would then become conscious of itself and, vice versa, the man would gain his materialized natural expression in the yucca. It is based on the nucleus, Man-Nature, that the dynamic of the Murui-Muinane people develops. For them, the difference between 'The History of Ourselves' and 'The History of the Punishment' is very clear.

'The History of Ourselves' is simply the original tale of how people appeared in the world, how they discovered their food, language, tools, everything that contributed to their culture. This history has many versions, depending upon the teller. By 'tradition' is meant the different ceremonial careers (rites) which exist within that tribe. During these rituals the men, in acts or words, change into contemporaries of the origin of the Murui-Muinane society.

'The History of the Punishment' differs basically from 'The History of Ourselves' in that it cannot be materialized in acts. It is a history of the 'Old Men', containing stories of those who failed under the power of the culture and of the society. They are stories to be forgotten. We have, for instance, the ritual of *Uik*,* which involved playing with a small rubber ball, but it happened that, through the special characteristics of the Murui-Muinane, they completely forgot this ritual, nor did they recall anything about the game, and the same thing happened with the ritual of the *Yarocamena*. It seems that the last time this ritual was performed was when the Murui-Muinane had to combat the inhuman exploitation of their people by the evil 'house of Arana' during the time of the rubber exploitation. But the ritual failed before the guns of the Peruvian Army: *Jayaga-Jitomaga* (history-light-punishment). It is another history which must be forgotten and which corresponded to a ritual. The social norm forbids remembrance of such tales, but individual anxiety perpetuates them, and so they are maintained in the memory of the Murui-Muinane people.

Statuary came into 'The Story of the Punishment' following a peculiar process. Since immemorial times the Witotoes' bitter enemy was the Carijona tribe of the Caribe group which, during the period of the Conquest, showed itself as warring and expansionist, and from whom the Witotoes received the name 'slaves', which in the language of the Carijonas was expressed as 'Ioto'. The

* Partly translated from *La Estatuaria Murui-Muinane* by Benjamin Yepez Ch., Banco de la Republica (Bogatá, 1982), pp. 17, 19 etc.

settlers and missionaries changed it into Jitoto, Witoto, Vitoto, which was finally used to name all the groups. At the times of their battles with the Carijona, the Witotoes used to devour their captives. Since then the statuary called *Janare* accompanied the ritual of supposed anthropophagy, while the other type of statuary, the *Janane*, had nothing to do with cannibal acts. Naturally, the 'civilized West' with its colonizing representatives immediately suppressed a tradition which they did not understand; as they felt contempt for and were ashamed of these manifestations, they declared the traditional manufacture of the statuary dishonourable, and so since about 1928 no more statuary has been made.

The existence of rituals among the Murui-Muinane is the form or the means by which this society establishes its mechanism of balance as a human nucleus, while at the same time it consolidates itself as a society in a specific medium which serves as a habitat. The ritual expressed verbally and in clear acts and manifestations its pedagogical intention on one hand, and the reorganization of the society on the other. Should the ritual derive from a conflict of occurrences (such as misfortune or illnesses) or if it is due to the chaotic breaking up of established social relations (inter-tribal conflicts), it is most important for the entire group to attempt immediately to obtain order by means of a revitalizing ritual. This type of ritual will not be repeated in a cycle, and it will not be one in which they may take refuge occasionally: its form and arrangements depend entirely on the circumstances of the conflict.

On the other hand, the Murui-Muinane have different 'ceremonial careers' expressed in rituals which are celebrated from time to time. Each one of these careers fulfils its purpose in the culture. These rituals depict 'The History of Ourselves' from different angles. However, their real intention is to make men conscious of the society in which they live and of their natural medium, and to make them at the same time contemporaries of the 'original beings', and also of the 'time of the beginning'. They are intended to demonstrate, as if for the very first time, the conflict of these men as social beings who must exist within an adverse natural surrounding. 'The History of Ourselves' is then put up as a model which should be followed but which also should be tested to ensure its validity and application. The ritual of the *Yuag* (dance of the fruits), is intended primarily to teach men knowledge of plants. It is commemorative and therefore must not always conform to well-established rules. Anybody can execute it at any time. It is performed, for instance, at the national celebration, on 20 July. The *Eufonegue* ritual is considered the most ceremonial ritual. To this ceremony, all the people of the Murui-Muinane will be invited. The distribution and interchange of food are an important part of the ritual.

In the *Aguira* the 'power' of an heir is passed on to the next heir of the tradition. These traditions are passed from father to son, and it is the father who determines who must be the heir of all his knowledge. If the person possessing the power dies before this can happen, his heir must include the name of his predecessor in the ceremonial career, and only then can he initiate the ritual circle of his own destiny. In order to understand completely the significance of the tradition of dreams and hallucinations, we must refer to the duality of the *Buinaima-Aima*.

The men of this tribe must know their role in their culture. However, not all men can occupy the highest position within the ceremonial career: only he who is elected by his own father as his worthy successor. The knowledge of tradition synthesizes within the man, the *Buinaima*, and so everyone in each of the ceremonial careers has his own specific *Buinaima*. All the *Buinaima* know each other in groups, and they are the multiple personifications which the yucca requires, as they depict the maximum expression of food. As a person, the *Buinaima* is the father of the community during the time of ritual, when he is the 'first being', the original one, he who taught and still teaches men to develop themselves in this world. He is the one who, generation after generation, assumes different manifestations. In the time of the ritual, the *Buinaima* becomes the nucleus that reorganizes and re-actualizes the origin of all things, of the activities of men, of their behaviour and of their balance, by accepting his examples of the social order.

But when he was elected as the heir of a tradition and when he is about to materialize his knowledge and is about to put it into practice, and to be the *Buinaima* he must be careful to maintain the rank he has acquired. He must descend to the depths of himself, and from there he must look at the spirit of his own tradition. To obtain this condition of hallucination, he will have to swallow a paste ('*ambil* of the mountains') which is derived from a tree, *ukaka*. In that condition, as he dives down into the depths of himself, he will go to where all the *Buinaimas* are assembled, and they will tell him, especially the *Buinaima* of his own personal tradition, that, 'He is already he himself and no other one, not different.' In that condition is then produced the connection between the new heir and the original *Buinaima*, who comes back to teach men how to live and how to be harmonious in the world.

To be a *Buinaima* means to be conscious that his own essence and his capacities are all derived from the yucca that all the implications of subsistence and culture are completed through the yucca. In addition, a *Buinaima* must understand the individual essence and the role he is to play among his own people.

To understand the duality of the Murui-Muinane mind – and, in fact, the coherent and relevant practice which is materialized in their statuary – we must understand that they select a certain timing. During the daily 'profane time', the Murui-Muinane men begin to determine themselves within their family nucleus or 'clan', as *'gente'*, meaning people of a certain natural or human nomination. This means to determine the specific clan producing the ritual. In terms of the society as such and referring to all the 'clans', the tribe is assumed to be divided during the 'holy time' of the ritual: Murui and Muinane, the significance of this division is:

*Murui*:      male,
             above,
             black,
             dances with leaves (ferns)
             bones
*Muinane*:    female
             below
             white
             dances with sticks (trunks)
             meat

In a Murui version of 'Our History', the men came out of the earth during the night, out of a 'hole of the people', and climbed and clawed, to open their eyes and to discover there were all sorts of foods. In the Muinane version, the men again came out of 'the hollow of the people' but they went down to the well, called *Uigogi*, and bathed in it. When they climbed out of the well, they realized that their navels had fallen off and had been changed into a giant snake, named *Agaro Nuyo*, which they finally killed and devoured. Some of them came out of 'the well of the people', which is a source of the River Uicocue, with the yellow feathers of the toucan, known today as Caraparand. The others came out of the well of the *gente*, which is situated at the source of the River Kotogue, meaning a guinea fowl-sandpiper, known now as the Igaraparaná. In every ritual mention is always made of the presence of people from above, namely the Murui, and the *gente* from below, the Muinane. These very limited notions of above and below represent earth and water.

The *Buinaima* is the person acting for the people. His teachings, his example and his knowledge are all those of the men who came out of the earth. The *Buinaisai* is 'he who has no personality', meaning he belongs to those 'original beings' who have their ancestors living 'below the water'. The former are real, the latter are not. It is difficult to comprehend *how* these two categories operate, because when they refer to the *Buinaima* they are very explicit, materializing him and inviting every three participants to follow his example.

The ceremonial careers derive from the 'original stem of the yucca', called *Jucitofe*. Also the first statues came out of this original stem, and their very existence is bound up with both *Buinaima* and *Buinaisai*.

The statuary known as *Janare* represents a human couple who are concerned with 'cannibalistic acts'. When a man 'descends to the very depths of himself' because he wishes to talk with the *Buinaimas*, he receives the order to make *Janare* statues. First, he must make the woman *Janare Buinano* and put her on the very bottom of the *maloca*, and then he must make *Janare Buinaima* the man, and he must put him at the entrance of the *maloca*. For this work the chief of the ceremonial career, being the chief of the dance, devised a dance in accordance with this tradition. In this dance, the master assumes the personality of the *Buinaima* of his own tradition, but he also mentions his conversation with *Jana Buinaima* who had permitted him to make the *Janare* statuary.

One of the important exercises in which the Indians of that village must take part while they are adolescents is to live for many days, without falling asleep. They continue this practice throughout their entire life. For these men the state of being asleep is thought to be the condition in which one is the most vulnerable to attacks by other men, especially by the *Aimas* (magicians) and by animals. Life is a constant fight, so one must always take good care.

The statuary of *Janare* has a meaning, in so far as it acquires the personality of its owner, who always will remain – with the help of a *Janare* – wide awake. So the presence of a *Janare* statue in the *maloca* guarantees protection to its owner, the *Buinaima*, before the aggression of the *brujo* (magician), called *Aima*. These *Aima* may act in two different manners: in song, questioning the knowledge of the *Buinaima*, and by acquiring the body of an animal, particularly the tiger. In the jungle night they may attack and produce illnesses, misfortunes, calamities. The *Buinaima* responds by replying to the songs and putting up other questions, also, in past times, assuming the body of the *Janares*. The former is the aggression/defence on the human plane the latter the aggression/defence on the magic plane. Under these conditions, the 'anthropophagique' act was situated on the magic level but was not surreal. And so the *Janare* would devour a threatening animal in the same manner in which *Buinaima* defends himself against the *Aima*. But in both cases a dramatic fight between life and death takes place.

The encounter between man and animal occurs daily. Animals are the Indians' immediate competitor in their natural surroundings, so the Murui-Muinane have assimilated extensive knowledge about animals, their habits and their preferences. Due to their knowledge, the Indians may either use the animals or hunt them – choosing, if they have a choice, to hunt them.

And so, in the last anthropophagical act, the *Aima* sacrificed his human condition in order to be considered an animal. He became not a human being but the personification of *Jana Buinaima*, the original being, father of the community. *Janare* was, in that case, the materialization of the 'original stick' of yuccas, called *Jucitofe*. It is finally the yucca (the food) which defends itself against the aggression of the animals, upon the non-magical territory of men. The dreams and the dreaming condition of the owner of the *maloca* will tell him that the annoyances which he is suffering, he and his people, are due to the spirits of *Janare* attacking him.

There is a long, rather involved story about the very first yucca, called Jucitofe.

Jucitofe did not stop growing. And out of the first stem of the initial yucca was fashioned the first status. All the many statues came out of the first stick of yucca. And so every dance begins with the wand of the stick of yucca. Nobody can begin a dance if he does not first produce the word power of the stick of yucca and ask if what he is undertaking can be done.

The Indians feel this is the same as when the white people pray to God. They say, when they start a dance, 'We will talk with the *Buinaima*, so that all will be well. And the *Buinaima* will give the name and show us how to make the dance. And of all the mentioned names, there is one single one which is "*Ru'be Buinaima*", and that name is the one joining all the other names ... and from there *Aime Jurama* took all that our grandfathers learned and they gave him all of the present-day customs and rituals.

Yucca (Yucca manioc, *La mandioca* in Spanish), of which the bread made is called *cazabe*, is the centre point of the entire Amazonian agriculture. After the burning has taken place before April and September, so that the rain will not wash the ashes from the earth, the indigene women are charged to plant and to attend to the *Yucca brava* (poisonous yucca, and other plants such as yota, mafafa and aji) and the *Yucca dolce* (sweet edible yucca), while the men cultivate, for example coca, Yaje, tobacco, barbasco, pina, chontadura and platano.

The growth of the poisonous yucca is the cultural watch regulating all the other activities of the *chagra* (*chacra*). This yucca ripens over a period of nine months. As it is collected, new seeds are being sown. After a batch of yucca has been sown, there will be sufficient to last eighteen months. After the *chacra* has been abandoned, people will come to gather new yucca, which will continue to grow freely, as there is no fencing around the plots. Recuperation of fertility in an abandoned *chacra* is rapid: very little weeding has been done and the Indians have no poisonous sprays. Their method of cross-sowing in apparently wild disorder of different species seems to protect the new plants against the multiplication of plagues. Yucca is their basic food, so it is easy to understand all the myths that glorify it. It is possible that the *Yucca Manihot Esculenta Crantz*,* of the family of the Euphorbiaceae, is endemic to the Amazon. Botanists say that there are still wild yuccas in Amazonia that have not been cultivated and may be subject to studies. The yucca, which is a shrub from 1.20 to four metres tall, nourishes with its roots men and animals and furnishes industry with carbohydrates.

The pupils of the *internado* in La Chorrera tell different tales. One describes 'the creation of the world'. When Moma began to create the world, he descended upon a very fine thread and could not find a place to stay. He thought deeply before he made some earth, but he could not use it, so he tried again, but only stones resulted. He thought once more and made stones and sand, and then he made mud. Moma was better pleased with this, but mud was still not good enough, so he mixed it with sand and stone, and out came a small seed, and the seed grew until it formed a jungle.

Another tale. The primitive Indians ate raw food,

* yucca = Maniok root

because they did not know how to create fire; but one day an Indian was rubbing some pieces of wood together and he made fire. It caught the dry leaves and roots and was soon out of control. Since then men have been condemned to work and to sow. Later, as a punishment for the fire, there came an inundation which destroyed the entire earth. All was lost: the bananas, the yucca, the *umari*. This punishment was in retribution also, for the sins of the Indians. Only one good Indian escaped the flood.

And when Moma looked at the world he had created, he saw that it was good and he rested. Later he made a man out of mud, but he was not satisfied, so he tried again, and the man came out as weak as a silk worm (*gusano*) so he destroyed it and made it again. After much thought he created a man who could not talk. He did not like his handiwork, so he destroyed it and made it again. The new creation was able to think, and this pleased Moma, because his creation resembled himself.

The Origin of Evil. When Usuma created beings, they all took the shape of small chickens and clung to him and tried to harm him. Usuma, using his word, which is '*jusinamuy*', shook them off, and for that reason his word is good, because he refused evil.

There were already many men on earth, but they did not respect one another. They wanted to be like Usuma, so Usuma punished them by changing them into stones, so that they would be the backbone of the world. And others he changed into snakes, sirens and evil spirits. And the good Indian remained on the shore, hoping that all this evil would finally stop.

And small animals drowned: ants, frogs, birds – all, as they came near the shore, left small sticks, feathers, bits of bark (*pelitos*) on the shore. The good Indian collected these and used them to make a boat, with a big roof, and he went into it himself, together with the remaining animals who had come to the shore, because they were all that survived the flood.

In another story, the creation of the world takes quite another form. (Here the memory of oriental reincarnation seems clearer!)

After the big flood, the *Buinaima* and his companion Buinino, and the seed of tobacco, and all the people were born. These people returned to the earth to hide there, but they were born again, this time with tails. As they came out of the hole in the earth, a huge wasp lay in wait for them and cut off their tails as they emerged during the night. Those that came out in daytime escaped with their tails and became monkeys. Those without tails went to bathe in a lake called *Uigobi*, which means 'troubled waters'. The hole in the earth and the lake are still in existence on the highest point of La Chorrera, which is called *adoki*. From there came six Indian tribes: Murui, Muinane, Bora, Andoque, Okaina and Nonuya. All these names were given by the *Buinaima*. This is the story told by Mitire Safiama, son of a Cacique who was a Murui, and his mother was a Huinane; he says that each chief gives another version telling about the creation of the world.

He states that Jusinamu was the real creator and master of the world, who gave it power, intelligence and wisdom, but he was assisted by his cousin who brought evil into the world. After the flood the *Buinaima* joined his cousin, Buinaino, and they named all the tribes. The name 'Witotoes' (or 'Huitotoes') is now being used as a collective term for these tribes.

The origin of various Witoto tribes does not seem to have been fully investigated. They are said to have lived in great isolation from other tribes, which makes it even more difficult to determine their religious beliefs and customs. They settled in small clans, or family groups, all united within their own *malocas*. It was known that they were among the most warlike tribes of Amazonia, that they were proud of their ancestry and rituals and that they were fiercely protective of the reputation of their god, Jusinamuy. Soon the Capuchin fathers began to understand that to each Witoto his family *maloca* represented an independent political entity, whose customs and ceremonies were transmitted faithfully from father to son. The family often came together for the most important rituals which constituted their religious celebrations, such as the feasts to celebrate coming of age, a patrimonial succession or the transmission of names. The chief would transmit his name to his eldest son at the end of such a feast, and at the same moment carved statues would be cast into the river. The missionary fathers found that basically many of the Witoto tribes were hospitable. Their *malocas* were open to any guest who might wish to share a meal of cassava (bread) with its inhabitants.

However, there seems to be no definite statement that the missionary fathers knew the meaning of the dances, which seems to point out that the Witotoes believed or believe in reincarnation. The fathers who tried and are trying to help their Indians in so many ways have difficulties in accepting that the Indians do not seem to plan for their future. Once they feel provided for, they may even forget to collect private or government subsidies, although this money may be obtained by the missionaries after long personal efforts. Another strange fact is that the Witotoes usually do not care well for their elderly parents, and yet the parents' love for their sons always seems very strong.

In their married life, the Witoto women give birth very easily, as they continue to work throughout pregnancy. They used to practice the *covado*, meaning that the father takes all the care while the woman gives birth. But the man used to fast strictly at the time when his son was born and abstain totally from eating meat until the son was about one year old. Marital fidelity is considered important for both partners, which would somewhat explain why at the evil time of Casa Arana women of the Witotoe tribe often preferred to be burnt alive rather than become an 'Indio-wife' of some white rubber-worker. At home, in the *maloca*, it is the woman who dedicates herself to the domestic work, but also to the agricultural labour in the *chacra*, while the men go hunting and fishing. When a member of the tribe dies, the chief informs the Indians through the *maguere*, a large drum. Then a number of persons gather, sitting and listening as the chief of the tribe relates the life of the deceased. Into the grave, usually within the *maloca*, there are always placed leaves of coca and *ambil*, tobacco.

# 8  The Mission's Work and Struggle

After the signing of the Salomón-Lozano Treaty in 1922 and its ratification in Rio de Janeiro in 1934, the Capuchin fathers from Catalonia began to open mission schools as quickly as they were able throughout the extensive and most inaccessible parts of Colombia entrusted to them. Two missionary fathers were nominated to work in the lower Caquetá.

Father Bartolome de Igualada bought a plot of land near La Pedrera, to build a school for Indian children near their homes, so that they would not have to travel too far. He and his colleague, Father Mateo de Pupiales, met with considerable opposition from the military authorities there, who had other intentions for the frontier, but despite this the fathers were able to have the foundations of the school officially approved.

In 1935 the first sisters of the order of Mother Laura arrived from Medellín to work at the new school. Sister Maria Esperanza, the mother superior, brought with her four Colombian sisters. At the same time, Father Lucas del Batet arrived from Leticia with building-materials. These Catalan Capuchin fathers had to find their own bricks and mortar to build schools for the Indian children.

In 1937 there were ninety-eighty pupils at the boarding-school of La Chorrera: sixty-three boys and thirty-five girls. Father Mateo brought more children, Indians from the Canjoas and Macuna tribes. Funds were lacking, and it was feared that the school might have to be closed, but somehow money was found, and in 1940 the new buildings were opened.

During those early years the fathers travelled up river as far as Miriti to bring children to the mission school. The Indians were happy for their children to go to La Pedrera until a malaria epidemic in which a boy called Esteban died. After his death, the children were no longer allowed to come. And so the population of La Pedrera dwindled, until by 1946 there were only 300 inhabitants. In 1947 the Ministry of War decided to build an airport at La Pedrera, which would bring great changes. Even so, it was decided to close the school in La Pedrera and to transfer it to Miriti. In January 1948 San Juan de Miriti was founded. In the long history of this mission, the school was frequently and repeatedly moved back and forth between La Pedrera and Miriti.

Another area of great importance in the history of the Province of the Catalan Capuchins in Colombia was Araracuara, where the Andokes (or Andoques) tribe lived.

As early as November 1905, two friars, Jacinto de Quito, and Santiago de Tucarres, visited the Andokes and other tribes of the upper Caquetá, before returning to a place called Tres Esquinas (Three Corners), at La Chorrera.

A commission was set up to establish a farming settlement and penal and prison settlement in a position near the Putumayo river. It was to be in a place from which escape would be practically impossible. In 1920 Father Estanislao de las Corts, then one of the commissioners, had selected a site at Caucayá which after the battles was named Puerto Leguizamo, to honour a dead soldier. The father began to initiate the building work, but the Peruvian government objected, and all building had to be suspended. Eventually a new location was found: the settlement would be built in Araracuara, but not at once. Many years passed before the clearing of Araracuara began. Finally, when the Salomón-Lozano treaty, formalizing the frontiers between Peru and Colombia, had been signed, a military garrison was established in Puerto Leguizamo.

Even in Araracuara the process of construction did not go smoothly. The Indians who had been brought in as labourers were assailed by a violent measles epidemic, which very nearly extinguished the Tanimuca and Letuama tribes. In September 1935 Father Bartolomé de Igualada visited the region during his trip from Puerto Limón to La Pedrera. Six years later, in 1941, the *corregimiento* (district) of Araracuara-Santander was created.

The Capuchin fathers wanted to open more missions for the benefit of the Indian tribes, but it was a long struggle. Now that they had virtually succeeded in eliminating illiteracy from so many tribes, they saw it as their main vocation to convert and baptize the Indians. They planned to train native catechists, some of whom would become priests and take their place in teaching their own tribes in their own language. All this would be possible only if there was one common language in which they could understand each other, and this language would have to be Spanish. So the fathers trained bilingual Indians to teach their fellow tribesmen.

The fruitful Putumayo valley, with its magnificent forests of giant trees, draped with lianas of great length, had been the envy of Peru for centuries but had now finally been allotted to Colombia. Travel into the Putumayo remained a laborious business: the only roads were those built by the Capuchin fathers; river travel took weeks and was fraught with danger.

At the beginning of the twentieth century, the first Concordat between the Papacy and the Colombian state, which gave considerable legal powers to the Catholic missionaries in Colombia, was still in full force, and the Capuchin fathers from Catalonia were responsible for large parts of Amazonia.

Colombians continued largely to neglect the Putumayo, the Amazonas and the Caquetá, the districts named after the three powerful streams. They were therefore only too pleased for the Capuchin missionaries to accept the responsibility of transforming the Indians into more or less 'civilized' Colombians. This was before anthropology was taught at Colombian universities, and differing opinions and viewpoints were introduced. Meanwhile, travel by air was soon to remove some of the main difficulties of communication with the rest of the country. Unfortunately, contact with the rest of the white man's world also brought vices and illnesses against which the Indians were often quite defenceless.

Topographically Colombia ranges from permanently snow-capped mountains to fully tropical swamps. Its flora and fauna have therefore attracted travellers and scientific researchers. Unfortunately, the Colombian government has often shown little understanding of the differences between the cool Andes and the hot, humid coast and interior, and their respective peoples.

In colonial times and when the land was called Nueva Granada (New Granada), the Spaniards and other immigrants of European descent preferred to live in the cooler, higher region of the Andes, formerly occupied by the Indians.

On the coast lived the African Negroes (and those of mixed ancestry, called Costeñans), originally brought to South America as slaves.

Air-conditioning has, of course, brought considerable changes to the area in recent years. The seat of government is in Bogotá, 2,660 metres above sea-level, and this is perhaps one of the reasons for neglect of the tropical Amazonian rain forest. Also, only 120 kilometres and one river bank of the Amazon river belong to Colombia. The situation is likely to improve, as tropical forestry is now taught as a science at Colombian universities. Also there is the organization INDERENA (the Institute of Renewable Natural Resources) which does all it can to protect the humid forests.

Still, it is strange to think that the area of Leticia, now invaluable to the Colombian economy because, since the loss of Panama, it is the only link between the Atlantic and Pacific oceans, did not belong legally to Colombia until 1930 (officially not until 1933). It did not attain its present importance until twenty years later, when it became the seat of the Apostolic Prefecture under the Catalan Capuchin Prelate Monsignor Marceliano Canyes, its first and only Prefect up to 1988.

Despite all the difficulties, which sometimes seemed insurmountable, the Capuchin fathers succeeded in opening more than forty-five mission stations and boarding-schools for the native children, giving them at least a basic education. In 1950, however, an event took place in Bogotá which was to bring profound changes to the fathers' activities.

That year Marcelino Canyes, the Capuchin father who called himself Marcelino de Castelvi de la Marca, died. He had studied in Bogotá with Paul Rivet and had exchanged knowledge and opinions with him. He had become a great scholar in Colombia, mainly by encouraging and collecting notes, observations and translations from missionary Capuchin fathers in the Amazon. He was a convinced follower of the Catalan Franciscan Ramón Lull, who wrote in Catalan and Arabic and came from Mallorca. Ramón Lull had originally taught that Arabs had a right to be converted in their own language, and had achieved the opening of the first institute on Mallorca for the study of foreign languages for missionary priests. Father Marcelino held the same conviction about the Indians, but this was no easy task in Amazonia, as there were so many unwritten languages and dialects. The Centre for Linguistic Studies (CILEAC), which Father Marcelino founded in Bogotá had been established with this aim in mind.

Father Marcelino was only forty-two years old when he died. There had been four brothers and two sisters in this remarkable Catalan family; all four brothers devoted their lives to the Church.

The youngest, Marcial, born in 1917, was shot in 1936, during the riots against the Catholic Church in Barcelona. The eldest, Marcos, born in 1901, became director of the Catalan Bible Foundation. He died in 1942. The two remaining brothers, Marcelino de Castelfels and Marceliano de Vilafranca, had left Barcelona in 1936, to work as Capuchin missionaries in Colombia. Although Father Marcelino had no official university degree, he was a great scholar and became a member of thirty-five scientific bodies. He died aged only forty-two, of meningitis, which his doctors said had been caused by continuous stress and overwork.

His brother, Marceliano de Vilafranca (it is customary that the Capuchins add the name of their native village to their first names), stricken with grief at the early death of his brilliant, much beloved brother, was summoned to Bogotá for a personal interview with the papal nuncio. At that time he had been in charge, first of the parish of Pasto, then in the Sibundoy Valley, and last in Florencia, capital of the Caquetá. After the death of his brother, he had been called to work in Leticia, in the full tropics. He wondered whether he was to be put in charge of CILEAC in Bogotá, the centre for Indian languages and their further investigations; also perhaps to establish a small mission museum of native stone art, centred around a collection of stone axes from the Caquetá. Or should this be opened in Leticia? Of the four Canyes brothers, Marceliano had always thought himself the one with the weakest religious vocation, but he alone had survived. But the nuncio told him that he was to be appointed the first Apostolic Prelate of the entire Amazon region in Colombia.

Leticia was still a very small city, growing slowly on the banks of the changing Amazon river, where the mighty stream is at its narrowest, at the south-eastern tip of Colombia. Colombians had not seemed to be much interested in expanding the city and the region, perhaps because of the extreme humidity of its climate. But now Leticia was to be made the spiritual centre of the entire region.

Between his arrival in Leticia and the following day, when he was to be appointed Apostolic Prelate, Father Marceliano may have spent a sleepless night. He would be confronted with vast problems in his new appointment. He was to be in charge of all the educational establishments in the region, a problem in itself. Not least among the difficulties he would face was regular communication with the other mission centres. Assailed with doubts about his ability to fulfil the great honour which awaited him, which he certainly could not refuse, Father Marceliano must have felt the spirit of his dead brother very close to him. Of course, he would accept the challenge, partly for love of his brother, and because he cared very much for the Indians, for their souls and for their minds, although both were difficult to influence.

Meanwhile, in Bogotá it had been decided that the Comisaria del Amazonas, as it was called, was in need of urgent attention. The Sibundoy Valley, where Father Marceliano had been a priest for twelve years, would be annexed to the diocese of Pasto. This would make the whole region one single prefecture of about a quarter of a million square kilometres.

During the ordination ceremony in Leticia, the new Prelate, although highly honoured by his appointment, did not feel entirely at peace with God. He admitted as much in his speech, saying how strange it seemed that it was he and not his saintly brother Marcelino who stood there. He had always known that God was perfect and infallible, choosing the best of humanity, in His eternal wisdom, for His service, but he was aware that, in his opinion and in his own case, the best – his three brothers – were dead, and he stood in their place. He accepted the task in front of him with great humility.

Father Marceliano, then thirty-nine years old, was an extremely handsome man, with an aquiline nose and fiery eyes, and looked all the more impressive robed in the Prelate's purple. Nevertheless, throughout the ceremony, in the full tropical heat, he felt weighing upon him the heavy burden of the future he faced. Then his eyes wandered among the congregation. They fixed first on Father Romualdo de Palma, a Jew, like Jesus, and now firm as a rock in his Catholicism. Like so many missionaries, he came from Palma de Mallorca, as his name indicated. And there was Father Cristobal de Torralba, the same age as himself, at that time the regular father superior of the district Amazonas. He might be willing to work permanently in La Chorrera. Were all of them prepared and willing to meet the double demands from Church and the Colombian government? The missionaries were supposed to be highly educated as teachers and yet ready to live amid the not so long ago cannibalistic Indians, and educate them. Yes, no doubt these Capuchin friars, his compatriots, were tremendously strong, but, how was he, their Prefect, to travel through mostly unexplored regions to the different mission schools?

Thus was Marceliano Canyes appointed Apostolic Prelate and Inspector of National Education of the Colombian Amazonia. The appointment was new and therefore without a budget or a salary. But the nuncio had told him clearly that he was expected to work very hard as the chief and spiritual leader of the entire province of Catalonia. Perhaps Spain's Catalonia would help him financially, considering the great honour of having a Catalan made a prelate in Latin America, in Colombia.

Before starting to visit the priests and brothers working widely separated in various places, all hoping to build mission schools and churches, he intended to visit his remaining family. He had not seen his beloved, now elderly, mother for sixteen years. In effect, she had lost all four of her sons to the service of the Catholic Church, and now he, her only surviving son, would not have many chances to visit her. He decided to travel to Europe as soon as he felt able to leave the missionaries to carry on their work without him. He left Father Romualdo de Palma in charge of the mission during his absence.

It was a blow when the budgets for his future work were finally given to him by the Colombian government and by the Vatican; they were very limited, and he could scarcely expect the usual charitable supplements from the residents of his parish. Everything would have to be built from scratch, with the slenderest of means.

One of the greatest and oldest difficulties besetting Leticia – and, indeed, the whole of the Amazon Valley – was the total lack of stone. There were no paved roads at the time, and all buildings had been made of wood, until the Capuchin fathers brought building-materials from Mañaos. Cement had to be brought by ship from Barranquilla. Bogotá was very far from Leticia, in so many ways. The distance between the two cities, and the condition of the roads and the rivers which separated them, made rapid travel well-nigh impossible until the advent of air transport. In the beginning, only the state airline, run by the Colombian Air Force, made the journey, using slow aeroplanes. Colombians, on the whole, did not like to live in Leticia. For instance, from 1966 to 1972 there were six different commissaries in succession in the region. So far there had been no tourism. However, there was stone in the Caquetá, and in the Putumayo, La Pedrera (meaning 'of stone'), together with Araracuara, had always supplied the entire region, as far as Brazil, with stone and stone implements.

Just before leaving for Europe, the prelate flew to La Pedrera by military plane, on his first apostolic trip. It was the only way to reach the mission stations with any speed, and he planned to visit his priests regularly so as to follow up and direct their missionary activities.

After visiting La Pedrera, the Prelate flew down to Miriti, spending Maundy Thursday and Good Friday there. The mission school had been moved from La Pedrera to Miriti but would be called back to La Pedrera, staffed by the sisters of Mother Laura. Up till then, very little had been done for the local tribes, who were Yucunas, Matopis and Andokes for the most part.

Monsignor Marceliano continued his journey to Araracuara, to the prison camp, opened there in 1940. The Prelate considered the camp to be the plague of the region, an inhuman place in which the then 1,700 prisoners were distributed in twelve camps. He decided that he would press the authorities for its closure.

There was at last a hydroplane air service between Bogotá and Leticia, but the Prefect's main problem of keeping regular contact with his Capuchin priests, friars and brothers remained. There was still no telephone, radio connections were quite unreliable, and there had been no air services to the mission stations, other than the military ones. Now at least there was a hydroplane. In cases of emergency, help could be sought from the Navy or the local army, provided they had no other problems to solve. For the moment, civilization meant that there was no longer a danger that the Indians might eat the missionaries, but it was not easy to encourage different tribes to collaborate in their own instruction. It seemed absolutely necessary, in order to train bilingual Indian leaders and teachers, to begin with the teaching of the very youngest children. For that they had to build more mission schools, *internados*, for primary education, which allowed the children to return to their parents only for holidays. Such a primary mission school had been working in La Chorrera for nearly twenty years, from 1933, and the Prelate hoped to persuade the Colombian government in Bogotá to lay down a landing-strip for aeroplanes, close by.

After returning from Europe, where he had spent only six weeks, the Prelate brought some money from Barcelona to Bogotá, where he was trying to influence the government. At first he used the funds from Barcelona, to last until the government capitulated and allocated a regional budget to his work in the prefecture of Leticia. He then bought a house in Bogotá, to be named the 'Casa de la Prefectura'. It would shelter the collection of ancient stone axes from the Caquetá Valley and also house his brother's precious legacy – his historical treasures, his library and, most importantly, his scientific notes, taken from the field-reports and studies of the Capuchin missionaries, which he had not had time to complete.

In Paris, Dr Paul Rivet wrote about this sadly unfinished work: 'I received from Father Marcelino Castelvi a long letter concerning the ethnographic and linguistic research he and his colleagues have undertaken into the little-known tribes of south-eastern Colombia.... Father Castelvi has begun to establish a list of existing documents about the different language families currently acknowledged.... This bibliography provides precious information about the unpublished works which exist in the mission....' Perhaps later scholars will continue the research cut short by Father Marcelino's early death.

In 1953 several important events occurred: in January, the parish of Leticia of 'Our Lady of Peace' was created, and the hospital of San Rafael was founded there.

The same month the 'Convenio de los Missiones' was signed, establishing an accord between Colombia and the Holy See. It confirmed that the missions were autonomous in their territories and especially in their work with the indigenous tribes and that their decisions did not have to be approved by the Senate or Congress.

Meanwhile, in Spain, the Association of Secular Missionaries (ASM) wanted to help, and in 1953 the first secular missionaries arrived in Colombia. Pilar Bea* came from Barcelona with the intention to help found the new mission boarding-school for indigene children at Puerto Nariño, giving her personal assistance and financial help.

In June 1956 the inter-urban telephone service was inaugurated in Leticia, with a network of some hundred telephones. This was a great advance in the modernization of Leticia and the surrounding area, but the problem of rapid and safe travel between the schools and the *internados* had still not been solved, and the modest needs of the fathers in these outlying stations remained unfulfilled.

---

* Pilar Bea returned to Colombia from Barcelona in 1986, wishing to see the *internado* and mission school at Puerto Nariño which she had helped to found. It was directed then by Father Romualdo de Palma. On her way back to Bogotá, from which she was to return to Spain, she was tragically killed in a plane crash. In that terrible accident of 1986, eighty-one persons were killed in the deep jungle near Puerto Nariño, where it was difficult to locate the remains. How many good intentions to continue helping the missions and the Indians went up in flames there!

# 9   Araracuara and the Andokes

The native Indians' right to free education, and the vital need to provide at least primary schooling for them, had long been acknowledged by many educated Colombians. Eventually, in 1960, within the Ministerio del Govierno was created the 'Division de los Asuntos Indigenos', charged to deal legally with all matters concerning the Indians: reservations, agricultural abilities, education, religious instruction, taxes, military service etc.

There was still a great deal of work for archaeologists and anthropologists in Amazonia. The entire area of the Caquetá had yet to be explored, and the petroglyphs along the banks of the river had not yet been deciphered, nor their age established. Crevaux had written about them after his visit in 1879, and in 1923 another French explorer, Tastevin, had sketched them. Monsignor Marceliano used to say good-humouredly that it was a pity the conqueror came to South America with Franciscan monks, instead of anthropologists. Many people felt that things might indeed be easier now if scientific research had preceded religion in the Amazon.

Meanwhile, the Capuchin fathers continued to build mission centres. In 1955 they built one at San Rafael de

Caraparana, and later that year the previously mentioned mission school and *internado* for indigene children at Puerto Nariño, which still exists. Floodwaters of the rising Amazon subsequently swept away the buildings where catechism classes were held.

In 1956 the Christian Lasallista brothers arrived in Leticia to run the Orellana secondary school. Also in 1956, the parish of St Teresita de la Chorrera was inaugurated, to include the boarding-school, which was now well established on the other side of the River Igaranaparaná, high up on a hill amid flourishing palm trees. Six years later, a census recorded 1,400 inhabitants in La Chorrera. The school then numbered 152 pupils, missionary work was progressing and at that time Father Christobal de Torralba, an energetic man, was the director. He had come to La Chorrera in 1943, accompanied by his saintly mother, and worked with Father Javier for some years.

After Monsignor Marceliano's visit, Father Javier de Barcelona became the first priest in Araracuara. This means that he was there while the prison-camp was still in use, in 1953 and 1954. As recorded, he was the priest who brought hens and seeds to La Chorrera, because the children had nothing to eat. Now, here in Araracuara, he observed how the prisoners were starved because any good food went to their guards. It was probably then that he decided to return later as a prison chaplain.

The second time the Prelate decided to visit this prison camp, in 1961, he was accompanied by Father Romualdo de Palma. In the central camp alone they saw four cemeteries. This was the place of the severest punishments. Wherever they went, they saw crosses made out of broken sticks, each one marking the spot where a prisoner was buried.

Not far from the track, there was a big corrugated iron shed, lined inside from the floor to ceiling with barbed wire. In this shed the worst offenders were locked up at night. Outside were some cages made out of tiles, within which there was hardly space for one man to stand up. In these cages men were locked up and left to suffer for weeks on end, exposed to the burning heat amid their own excrement.

Fearful that conditions might become publicized in some way or other and that the authorities would seek to make changes, the managers of the penitentiary opposed visits by dealers and travellers from the interior of the country.

The situation at the camp soon began to attract attention. In 1956 a number of Jesuit fathers came to visit from Bogotá. Two years later Father Norberto del Prat settled in nearby El Chuquio where later the airport would be situated.

On 6 July 1959 Dr Alberto Lleras, then President of Colombia, announced for the first time his intention of closing the prison camp of Araracuara. Following this announcement, a mission was established there, and the new road, forty-seven kilometres in length, between Miriti and Santa Isabel in the Caquetá, was completed.

Despite the presidential announcement that the prison was to be closed, the cruel treatment of the prisoners continued. Even a serious accident in 1962 did nothing to precipitate its final closure.

The thunderstorms near Araracuara are said to be so violent that they are engraved in the myths of the Andokes. During such a storm, a river launch, carrying the General Director of Prisons, a visiting architect, the local prison director and the doctor, was drawn into the rapids of Angosturas and completely destroyed. All aboard were killed. The prison of Araracuara, however, did not close until ten years later.

In 1960 the sub-parish of Puerto Nariño had been inaugurated and provided a radio-telephone service between all the Capuchin missions in Colombia.

The sisters of Mother Laura from Medellín came back from Miriti to teach in La Pedrera. The Prelate had decided that the *internado* would be sited on the hill. He had sent Father Javier de Barcelona as a chaplain; he was to return in 1970 and establish his residence.

The Andokes who lived near Araracuara were known as 'the people of the axe', because of the fine stone axes they made. This tribe, like the Witotoes, had its own language but showed certain cultural similarities with neighbouring tribes. In 1915 there were still over 10,000 Andokes, but the Casa Arana almost wiped them out.

About 1,300 of them lived in scattered dwellings near Araracuara. Close to river banks or near swamps, the Andokes built their houses on stilts. They were good farmers: the men cleared and burned the fields, like the Witotoes, and planted coca and tobacco; the women sowed and harvested. Some members of this tribe were apparently found years before by the Colombian Army, who sent them to Iquitos in Peru, where they continued to live. One day twenty-one Witotoes appeared in the area where the Yaki river joins the Caquetá. They had fled from the Casa Arana many years before. They settled near the mission station and seemed quite eager to learn as much as they could from the Capuchin fathers there.

In April 1970 Father Romualdo came for a visit, and in June of that year Father Javier de Barcelona decided to move to the nearby village of Santander. Sadly, in December the same year he fell seriously ill and was flown to Bogotá, where he died on Christmas Day, a great loss to the mission. Unlike so many others, he was never discouraged by the Indians, because he loved them.

Strangely, a year later a fire completely destroyed the house where Father Javier had lived, as well as the district administrative offices and the post office. The fire appeared to have started by accident, but a few months earlier an unknown arsonist had deliberately set fire to

Father Javier's living-room and study, burning all the valuable files and documents the father had collected there.

In 1972, after three years of hesitation and neglect, the Ministry of Justice finally closed the prison camp and farming settlement in Araracuara. It gave the properties to the Ministry of Agriculture, which planned to create a farming and stockbreeding centre there.

The most significant change in relation to the education of the Amazon Indians took place in 1975, when Marceliano Canyes, Prelate of the entire vast region, handed over the responsibility concerning general urban education in Amazonia to the Colombian government. By that time illiteracy had been largely eradicated in the Colombian part of the Amazon, and Catholicism seemed firmly rooted and accepted by the great majority of the native population.

But the Colombian government was faced with major educational and scientific problems and needed foreign help. It was decided to establish a Research Centre for Amazonia, as a joint scientific farming venture between the Colombian and the Dutch governments. The Dutch Government would provide technical assistance for nine years, as well as financial support of some 80 million pesos, while Colombia provided land and buildings. The main aim of this venture was to carry out research into the best uses of the sensitive Amazon soil. Experiments had confirmed that, while the first harvest in that part of Amazonia was always splendid, the next would be meagre, and the third non-existent. This was the reason why the Indians used to move their *chacras* (small farms) regularly, burning another part of the rain forest and clearing another area.

Research was also to be carried out into practical ways to protect and breed giant turtles.

In August 1977 the Stockbreeding and Farming

Co-operative of Araracuara (Cooperación Agropecuaria de Araracuara) was founded, with an office in Bogotá. It had received and accepted the land, buildings and stock from the Colombian Agricultural Ministry. Everybody there and in Araracuara worked feverishly, but strangely enough nobody thought of consulting the Capuchin mission fathers, with their long years of regional experience.

In 1977 an Anglo-Colombian scientific expedition into the Amazon was also planned. It seemed to be the beginning of a productive relationship between Colombia and Europe.

A new *internado* for Indian children was opened in Araracuara by Father José Maria Clarassó from Catalonia. This was built on the site of the old prison camp. Father Clarassó was director of the mission school from 1975 to 1978. Monsignor Marceliano came to this mission station again in April 1977 to give his blessings to the school. Father Clarassó wanted to push his Indians along to a higher degree of education. And so he did, until Monsignor Canyes came for his brief visit.

All through the thirty-five years of the Prelate's career in Leticia and as Prefect of an enormous territory, there had been, somewhat suppressed, a difference of fundamental opinions between him and his priests, although they were obliged to obey him. Some of the fathers and friars had been much in favour of pushing the Indians along, even to take university degrees, provided they showed enough initiative, endurance and a real desire to study. Monsignor Marceliano was much more hesitant in this respect, for several reasons. At the end of his own career, he said that those who had worked for many years exclusively with the Indians often lost their drive and grew weary. Other priests, such as Father Juan A. Font, a much younger priest, defended the theory that the Indians had their own religions and that it was up to the priests and professors to study their background more carefully and then adapt their teachings in accordance with each tribe's belief. Time has shown that both were right to a certain point.

The 130 Andokes still living below Araracuara, near the mouth of the Anduche river, which flows into the Caquetá, continued quite stubbornly to tap the few remaining rubber trees, in the most miserable living-conditions. And so, from February to May and from August to September, these Andoke families worked on the few rubber trees which still grew at the heads of the Yavari, Mesai and Cuñare rivers. During the intervening periods, they worked in the fields and celebrated their ritual dances. Since the establishment of the mission station and chapel, they had started to build *malocas* again, instead of houses on stilts, as they no longer feared that the level of the rivers would rise so dangerously high.

After the opening of the *internado*, the Capuchin fathers noticed that the Andokes had become rather modern, with their wristwatches and their aluminium cooking-pots. Having the insight to look more closely, the fathers also sensed that the Andokes' ancestral mythical world was still intensely alive within them. Outwardly they dressed like any peasants, but inwardly they clung to their traditional beliefs. In order to understand them better, the Capuchin fathers took a great interest in their legends. Some of these showed evidence of the strong impression made upon the Andokes by the arrival of white people among them.

According to the Andokes' mythology, the entire region around Araracuara had been inhabited, before they themselves arrived there, by people of extraordinary power called Jonake. The Jonake gave names to all the rivers and to the important places; it was they who had carved the mysterious petroglyphs on the river banks.*

---

* In 1975 the anthropologist Elizabeth H.R. von Hildebrand chalked all the

They tried, in vain, to go up the Anduche river, a tributary of the Caquetá. Those who survived this expedition returned to live in La Pedrera. The Jonakes' emblem was the white heron (*garza*).

The region inhabited by the Andokes was not penetrated by white men until comparatively recently. Colonization of the area east of the River Caraparaná, between the Putumayo and the Caquetá, did not start until 1830, when the first Colombian rubber-traders came. They established themselves primarily on the shores of the Caraparaná and Igaraparaná rivers, at the source of the Cahunari, and in the area between the Igaraparaná and the Caquetá. However, barter for 'white goods' dates from earlier times. Iron axes came from Brazil to the neighbours of the Andokes at the Cahunairi river. The older Andokes used to say that, long before Casa Arana, steel axes came to the lands of the Andokes, and even sewing-machines, which were exchanged for Indian women and children.

The steel axe was a more efficient substitute for the stone axe and saved a great deal of time. But the stone axe had held a mythical significance of some importance among the Andokes. As there was so little stone in the whole of the Amazon region, they had to use stone axes made by earlier civilizations, which they found buried at a depth of several metres. The very way in which these axes were then extracted was sufficient to provide a ritual: they could be excavated only by unmarried youths, and then only after consultation with the magic men. The rarity of stone in the Amazon valley became the basis of inter-tribal exchange and commerce which included practically all the

---

petroglyphs along the River Caquetá on the shores between La Pedrera and Araracuara. She points out in her long article published in *Revista Colombiana de Antropolgia* (Volume XIX, 1975), that these stone engravings are not always visible; only when the water-level drops.

tribes in the region, and most of the goods in which they specialized.

The Yukara were famed for their pottery; the Boras for mats, arm-bindings and giant bugles; the Carijons for their poisons, and with the Andokes they also bartered domesticated dogs in exchange for people; the Witotoes specialized in hammocks. All these goods were passed from one tribe to another through personal contacts and the giving of presents.

An Andoke narrator tells the story of 'The Birth of Goods':

This is the story of the heron 'Egreta Alba'. The daughter of the heron chief of the centre of the earth was the mother of them all, mother of the axe and mother of the Andokes. She had no husband, and no man had touched her. The sons she bore came from herself alone, from her thoughts, from her history. She gave birth because there was nothing. I shall tell you the story.

Her father and her brother had their own living-quarters (*mambeadero*). She lived next to them, imprisoned in her *maloca*, so that nobody could see her. One day she gave birth to a dog, and her brother asked her: 'What is it that to which you have given birth?' She answered: 'It is a brother to you, to keep you company.' Then she gave birth to a chain to tie the dog up, then an iron axe, then a machete, then a cooking-pot, a plate, a knife, clothes, a hammock ...

As all this was happening, the news spread to the mouth of the River Caquetá, where the heron chief of the estuary Okayofi said to his people: 'The daughter of the heron of the centre is producing tools. I must go and bring them back here.' He travelled to the centre with five people, and with mules. They arrived at the centre, and he went straight into the *maloca* where she lived. While he was talking, to her, her father and her brother remained silent. She told him all that she had borne, and she began on that day to write for the first time, to note all the things down. They packed the work tools into a cardboard box, and the heron of the estuary carried these and other things to the estuary. There were four boxes of tools left in the centre, because they could not carry them all. They were left without a mother, because the Mother of the Axe had also gone to the estuary.

After a few days, the brother said to his father in the men's quarters, 'Father, I am going to see my sister.' The Father said: 'Good, take some coca and some *ambil* [extract of tobacco] with you as gifts.'

The daughter of the heron of the centre saw her brother coming through a window and told the captain heron of the estuary: 'Here comes my brother.' Her brother came in and the heron of the estuary asked him: 'Where are you going, brother-in-law?'

'Nowhere,' replied the heron, 'I am happy and that is why I have come to see my sister and to fetch some work tools.'

'There are no tools,' said the heron of the estuary.

'What do you mean,' said the sister. 'There must be some. It is written here,' and she took the paper on which she had made a note of all her tools. Then she looked into the box where the tools were kept. Underneath all the others, they found the axe. 'Is this what you need?' she asked. 'Yes,' he replied. 'How many do you need?' 'Only one,' he said, 'no more.' 'I need a needle.' 'There are no needles,' said the heron captain. Again she looked at the paper and found it written down. She began to search: 'How many do you need?' 'Only one,' he said, 'no more. And I need a fishing hook.' She showed him a big hook. 'No' he said, 'I do not live on the shores of a big lake, to use such a hook. I need a hook as small as the teeth of the ant, so that I can fish with the eggs of the ant.' 'What else do you want?' 'I want no more. I am leaving now.'

The heron returned to the centre and told his father all he had done. 'Good, my son,' said the father, 'now we know how to get the tools, we can get more.' 'Father, here is the axe, the hook, the thread, the needle,' said the son. 'I have brought no more.' 'Very good,' said the father. Then they began to hear all things, in all places, as if for the first time. People talked and said that the axe had come to the heron of the centre. So the 'people of the axe' received the steel axe before any others. The other tribes said: 'Let us go to the centre to search for the tools, let us find the people of the axe, who can fell the biggest trees and clear the largest fields. Let us find the Andokes, because the mother of all the tools was an Andoke!'

The story is said to be allegorical. The heron of the estuary is the white man, indicated by the presence of the mules. All the references to windows, rooms and boxes describe the white man's houses, as the *malocas* of the Andokes have neither windows nor separate rooms. The heron of the estuary does not enter into a relationship with the father or the brother of the Andoke heron; in other words, he does not observe the rules of Andoke etiquette according to which a visiting man would come to the *mambeodero*, reserved for men. The reason why the chief of the centre, as well as his son, remained so silent when the captain of the estuary arrived was that they did not know how to make more tools. If the heron chief of the estuary had not taken the tools so that they could reproduce, and the mother of all the tools as his wife, there would have been no more tools. The father and the mother knew this. The heron of the mouth of the river, the white man, is represented as both a friend and an enemy: as friend because of the relationship, but an enemy because he denies that the tools exist. In one version of the story he offers perfume and a hat, instead of tools. The perfume is said to represent influenza and the hat headaches. Because of this, the heron of the centre does not accept them. The diseases brought by the white man destroyed whole Indian villages. The goods he brought could be used in the Indian's world, but the Indian was selective in those he accepted and those he refused.

The myth about the white man and his tools is constructed formally around one of the Andokes' fundamental spatial structures: the source of the river, where the sun rises and hides away; the mouth of the river, the estuary, where the river ends and the sun sets; the centre is the middle or the current of the river. The creator of all things is the magician, who sees all and knows all.

For some more modern Andokes, the myth portrayed a simple relationship: The heron of the estuary represented the Brazilians; the heron of the centre the Andokes, and the heron of the source was the Colombians. This interpretation probably derives from the fact that the current of the mouth of the River Caquetá flows towards Brazil, while the Colombians, at the end of last century, colonized the upper Caquetá, nearer the source. The fact that the heron of the source is mentioned only briefly could be an indication of the limited commerce with the Colombians.

# 10 The Birth of Conservation

The inaugural flight of Avianca's weekly service from Leticia to Bogotá, which took place in August 1960, had been a great step forward, in particular for the mission. This same year also saw Capuchin Father Cristobal Torralba begin work as parish priest of the new church of Santa Cecilia in Bogotá. Father Cristobal could now take care of his friend, the Apostolic Prefect, Marceliano Canyes, badly burnt in a boating accident on his way to Puerto Leguizamo. Monsignor Marceliano recovered after some months' nursing and returned to Leticia. At the time of the accident, he had been travelling with an American of Greek origin, called Mike Tsalickis, and they had struck up a friendship which was to prove in some ways very useful for the mission.

Mike was married to a pretty but poor girl from Leticia, and they had Colombian children. He owned an aeroplane and flew regularly between Leticia and Miami, his home town. This was at a time when air travel was almost unknown within Colombia, and any trips to the interior had to be made laboriously by boat. Mike's plane furnished the Prefect with the means of reaching his missions quickly and easily. It also made the Prelate somewhat unpopular with the foreign naturalists who followed Mike to Miami, where he owned a private zoo. But in Colombia, in Leticia, Mike was known for being charitable and supporting the hospital.

At that time, there were no nature conservation laws in Colombia, and the rich store of tropical flora and fauna was being ruthlessly depleted by export from Leticia and Barranquilla. It was a complicated situation; there was no official organization to enforce such laws, even if they had existed. The naturalists came to realize that Colombia's precious natural resources, especially those of the ecologically fragile Amazon, the habitat of the Indians, were threatened with destruction. The mission, guided by its Prelate, had little consideration for the animals exported to the United States, and supported the Indians who were able to earn some money from time to time by working for Tsalickis.

More than 300,000 Indians remained in Colombia, needing to be fed and educated if they were to live in the white man's world. Education in Colombia was an expensive business, but it was out of the question for such a large number of natives to persevere in a state of savagery. There were many problems to be overcome, as the Indians' habitat, where they farmed their small plots of land, was progressively diminishing; their livelihood was being taken away from them and not replaced. In addition to this, Indians were not accustomed to planning ahead; if they could eat today and sit in the sun, they would not worry about tomorrow. They had never used money. All these requirements of life in a modern world had to be instilled in their minds, so as to satisfy modern demands.

The Capuchin missionaries had tried for years to help these Indians to overcome their many problems and had come a long way. Theirs was an illustrious history. So many missionaries had devoted their lives to converting natives in far-away lands. Ramón Lull from Mallorca had preached to the Arabs; while the two Serrer fathers, Junipera and Juan de Santa Gertrudis, had travelled to America. The Jesuit father Pedro Claver, who had heard in a Mallorcan convent about the misery of the Negro slaves coming from Africa, had come all the way to the Colombian Cartagena of the Indies to baptize some 300,000 of them, often bribing them with small gifts to allow themselves to be baptized; he was later criticized for this. The German Franciscan nuns, the French Vincentina sisters and the Colombian congregation of the Sisters of Mother Laura, from Medellín, had all tried to help the Franciscan and Capuchin fathers, wherever they were needed, especially in their work for the Indians.

It was arguable whether these missionaries and trained teachers were also morally obliged to act as social workers. In any case, their great efforts and sacrifices seemed never to satisfy the government in Bogotá.

As research continued and the country began to be more accessible by air, a strong anti-Catholic mood began to pervade the Colombian anthropological society in Bogotá which had sprung from Dr Paul Rivet's love for the Indians. Some wanted to ensure that the Concordat between the Holy See and the Colombian government, which empowered the religious missionaries to baptize, educate and rule the Indians without being subject to the authority of Congress, was not to be renewed, or at least not in that form. In the past, Colombia had been a strictly Catholic country, but now there was a drive to introduce civil divorce and also family planning. Many articles, books and even theses were written by some of the younger anthropologists, criticizing the Capuchins' work, especially their establishment of the mission schools for Indian children, which some Colombian anthropologists claimed to be against the country's constitution.

The fact that these schools were originally established as orphanages for indigenous children whose parents had been cruelly killed by the lawless rubber-growers was forgotten. Forgotten also was the important fact that nobody in Colombia at that time had been particularly interested in the unrewarding and expensive task of educating these native children. The young anthropologists were more interested in developing their own ideas and furthering their research into the different small tribes' ways of life. Their minds were set on university degrees, and they were convinced that the missionaries' work had contributed only to the so-called 'deculturization' of these Indians, when they were taught to read and write Spanish and cover their nakedness.

The Capuchin missionaries were usually too busy to read all the material published in Bogotá attacking their efforts. Consequently the *internados* were often criticized and rarely given financial support, but no better way of education was established in the vast and often still unexplored Amazonian regions.

Eventually it was decided that 'the educational programme and institutional structure of the Amazon, including *internados* for the Indians, were to be brought in line with the official programme of primary education for the country as a whole.' This decree of the Ministry of Education in 1963 stated that primary education should provide all the Indians of the region with sufficient instruction for them to be able to continue secondary studies in any institution in Colombia.*

This was all very well, but as higher education was still very costly, scarcely ten per cent of the Indians were able to fulfil their wish to live in white surroundings and also continue their studies. For this reason, the Prefect and his priests planned to open a centre, although it would take many years to establish because of the inevitable difficulty

* See page 312 – 'Revista Colombiana de Anthropologia', 1975, Bogatá Vol. XVIII.

of obtaining government funds promised to the Prelate. The fathers were constantly frustrated by the problem of feeding their hungry pupils, let alone providing facilities to educate them.

Unimpeded by any such formalities, Mike Tsalickis began to do big business in Leticia. Sometimes he was generous with the Indians who worked for him, but they had to earn their pay. As requested, they brought him and his dealers live animals, specially monkeys, for his zoo in Florida, because he 'loved animals' – or so he said! They brought him skins of the most valuable black caymans (a species of alligator). At that time Tsalickis must have thought it would do no harm to have the protection – official or unofficial – of the Catholic Church in Leticia, and also of the US government. There was a great deal of confidential talk between himself and Monsignor Marceliano Canyes. In about 1967 a United States consulate was established in Leticia, and Tsalickis was appointed its first honorary consul.

In August 1967 the Hospital of San Rafael was officially opened in Leticia, a sign of progress. The population of Leticia was then about 17,000, including some 4,000 living in the surrounding area.

1967 saw also the foundation of the ATA airline, which then owned one small plane, a Cessna. George Tsalickis, Mike's brother, came from Miami to act as a pilot for the Prelate on his visits to the far-flung mission stations, and also to transport skins for delivery and monkeys to the United States for experiments, as well as beautiful tropical birds for pet-shops there. As yet, such exports were perfectly legal, not forbidden by Colombian laws.

Many of the Indians in Amazonia still lived in the hope of remaining undisturbed in their mysterious rain forests. They, whose nature it was to care for animals and plants, from which they often took their family names and around which they wove their legends, were paid by Mike Tsalickis to bring animals, alive or dead, to the white dealers of Leticia. Later on it was said that one hundred traders reported to Tsalickis. His dealings were protected by his consular status. He bought an island in the Amazon river called Santa Sofia, later to be called Isla de los Monos, Island of the Monkeys, so that he could guarantee the export of sufficient monkeys for experiments in the USA. And so monkeys living free on this island were caught and shipped in rough crates or boxes to Miami, USA. As long as the animals seemed to be in good health when leaving Colombia, there was nothing to prevent their entry into the USA, so, veterinary surgeons in Leticia would sign certificates confirming the animals' health, usually without even having seen them. Small parrots destined for sale in American pet-shops often took diseases with them which were contagious to man and domestic animals. This traffic continued for years.

Eventually, the authorities realized what was happening and decided to do something to prevent further imports on these lines. Also, animal protection societies in Europe and in the United States set to work.

Monkeys especially would be exported to Tsalickis' Tarpon Zoo. Provided they had survived the journey from Leticia, they would be sent to laboratories, pet-shops and anywhere else willing to pay for them. Business boomed. But the animals frequently suffered and died.

Leticia continued to develop steadily. In 1968 it was visited by the President of the Republic, Dr Carlos Restrepo. At that time, the mayor of Leticia was a woman, Dr Ines de Zapata. In May 1968 the Hogar Virgin de la Paz, a home for abandoned children, was founded. And in July of that year a whole planeload of cattle arrived, to replace stock in a region threatened with foot-and-mouth disease. The Ministry of Agriculture took charge of farming and cattle-raising there.

In the middle of that year, SENA,* an organization for the teaching of manual skills, came to Leticia. INDERENA,** which had been founded in 1960/61, for the conservation of natural resources, began to formulate laws after 1968 and to take action against the export of animals without a special licence and bona fide health certificates, controlled by INDERENA.

The Indian was now thoroughly confused. Before, he had been paid, albeit poorly, for bringing animals and skins to the dealers. Suddenly he was told that INDERENA had the right to punish him for catching birds, for dealing in animals and in skins and even for felling trees on his own farm, his *chacra*. In addition to this, fishing was soon to be restricted. What was he going to eat now? Years before, Leticia had been a small village with wooden houses, many trees along the streets and in gardens, and also a large park in front of the parish church, with a small lake with giant Victoria Regis plants, all surrounded by palm trees, framing a view of the mighty Amazon river. Now it was dusty and very noisy and there were too many cars for its few paved streets. It had always been hot, but not always so dirty. The harbour had been romantic and there had been an abundance of fish for sale; now fish were scarce, and when the waters of the Amazon rose, they flowed away from Colombia to Brazil. What would the Indian children eat? And what about the tourists who might come and buy native handicrafts?

Meanwhile Tsalickis pocketed the money he had earned exporting animals and skins. The hungry Indians were chewing more and more coca leaves, basis of the drug cocaine.

---

* SENA: National service for apprenticeship etc.

** INDERENA: Instituto de Defensa de Recursos Naturales Renovables. All national parks in Colombia must be determined, installed etc and permanently protected. Tourists are allowed in certain sections only.

In 1970, just before the annual cycle of floods in the Amazon, which forced the river-dwellers to leave their villages for higher ground, the Olympico Amazonas club was formed. The Prefecture supported it, hoping it would lead to the establishment of regular tournaments in the six districts of the Commissariat. The Indians had always been skilful and enthusiastic ball-players, traditionally using a ball made of rubber. The old game was to move the ball around by touching it only with shoulders and knees, never with the hands; now they played according to modern rules.

Leticia seemed to be progressing in leaps and bounds. On 6 August 1971 another President of Colombia, the Conservative Dr Michael Pastrana Borrera, arrived to meet the President of Brazil, who was accompanied by a staff of botanists and zoologists, to sign an agreement relating to the ratification of the frontiers. During this important visit, the new administrative centre of the district and a main branch of the state bank, the Banco de la Republica, were inaugurated, as well as a new electrical plant.

Meanwhile, in Bogotá and also in the Andean mountains, a parish priest, Father Joaquín Salcedo, founded, with the financial support of UNESCO, in a village named Sutatenza, an educational centre, called Acción Cultural Popular (ACPO), which was the first organized radiophonic school in South America, especially for farmers (campesinos). ACPO received as a donation from Pope Paul VI, on the occasion of his visit to Bogotá, the largest radio station then existing in South America. Salcedo, later Bishop of Bogotá, was firmly convinced that the biggest problem in Colombia, perhaps in all of South America, was the illiteracy of the population. His personal aim was to eradicate this completely.

Small radio schools, often organized on private farms, later in village schoolrooms, usually in inaccessible regions where it was difficult for the children to travel to the next school, could be used by a farmer or his wife to teach children or old people how to read and write Spanish. Later on, courses for farmers would be given. These courses were as a rule very early in the mornings, usually from 5 a.m. to 6 a.m., and with them came a free supply of *cuadernos* (pamphlets and paper). In the long run, the local region and then the entire country would perhaps benefit, as the voice of the radio could reach out to far-flung areas and bring them more information and education, especially to places where there were no schools. When Monsignor Salcedo received a private donation, earmarked to bring ACPO to the Indians on the Amazon, he accepted the offer.

While the mission stations, schools and *internados* were being established by the Capuchin fathers, Bishop Joaquín Salcedo agreed to undertake a special campaign to educate the Indians in regions of the Amazon, Putumayo and Caquetá, funds permitting. Bilingual Indian leaders would be trained to enable them to translate the radio programmes into their own tribal language. First they would spend some months learning about Andean agriculture in Sutatenza. Later, the graduated 'leaders' would receive a small salary for teaching their own tribes in the radio schools and thus helping them to attain a basic educational level.

Soon there was some justifiable criticism, also by the Capuchin fathers, that these ACPO programmes were too general and not at all adapted to the particular problems of the Indians in the Amazon, the Putumayo and the Caquetá. There was certainly a degree of truth in this, but at least some hope of education was offered in an otherwise neglected area, and also, very importantly, the Indians were brought into contact, on an equal footing, with other Colombians. In 1970 ACPO opened offices in Leticia, under the direction of Domingo Cabrero, who was able to visit in person the various radio schools in the region. Then in 1972 a literacy programme was started for the adults living in farming settlements in Amazonia.

At first ACPO was a great success, especially with the natives, who enjoyed travelling to Sutatenza, with air transport provided free both ways, often returning as 'leaders'. However, the idea lost considerable impetus when Monsignor Salcedo left Colombia to live in Florida, although he retained an indirect interest in the management of ACPO. He had often written in favour of family planning in his weekly paper, *El Campesino*, and perhaps because of this had forfeited the support of Rome. Later he resigned totally, for health reasons. For some years more he retained his interest in the management of ACPO, but the withdrawal of his forceful and very stimulating personality from Colombia was a source of great regret to his collaborators. Ever since the late sixties, the question of birth control had caused bitter controversy in Colombia, while abandoned street urchins in Bogotá, called *gamines*, continued to live precariously on the proceeds of petty crime.

Meanwhile the Capuchin fathers continued their work and would always help the Indians to get to Sutatenza safely, at least until SENA took care of these trips and also paid for them.

In 1968 Victor Daniel Bonilla published a very critical book in Bogotá under the title *Siervos de Dios y Amos de Indios* (*Servants of God and Masters of Indians*), in which he attacked the order of the Catalan Capuchins rather violently for having leased out to Indians in and around Sibundoy land that really belonged to them. The fathers replied in 1970 in an article published by the paper *Cultura Narinense*. In 1946 the Capuchin fathers had given the land to the Indians, but Bonilla's book became the indirect cause why in 1971 the Capuchins withdrew from Sibundoy and the Redemptorist fathers of the Apostolic

Vicariate of Pasto took over the entire Mission of Sibundoy, situated some sixty kilometres from Pasto.

The history is as follows: In 1621 an *oidor* (judge) by the name of Quiñones, had created the *resguardo* (reservation) by order of the Spanish Crown, so that all the land really belonged to the Indians. In 1700 an Indian chief, Cacique Carlos Tamabiory, made his last will with the purpose of reaffirming the rights of the Indians over the entire *resguardos* of Sibundoy and Aponte. But the white authorities, the *encomenderos*, did not recognize the rights of the Indians over the *resguardos*. On the other hand, law No. 89 of 1890 defined as 'savages' all those Indians who had no *resguarda*, which meant that these Indians were simply considered nomads, although they were usually living in hamlets.

In 1902 all privileges granted by the Colombian government to missionaries were augmented. In 1904 the Prefectura Apostolica of the Caquetá and the Putumayo was created. It was the function of the Capuchin fathers to serve the Indians of the valley of Sibundoy and also those of the medium and lower Putumayo, such as the Sionas, Coreguajes, Macaguajes and Witotoes. The missionaries were fully authorized to create *haciendas* (farms) on uncultivated land, given to them by the nation, and there were instructions that all civil persons, selected to fulfil different positions of the government, had to be completely agreeable to the apostolic delegate. As these *resguardos* were not recognized by the civil authorities, in 1916 the repartition of the land was to start. So the Capuchin fathers named a lawyer to defend the interests of the Indians. But during the prescribed period of eight months the old titles could not be found, so the Indians got desperate and started selling the improvements they had made on their land. It was in order to stop this that the Capuchin fathers put into practice the programme, much attacked by Bonilla, of letting the land out for seven and twenty-one years. According to Nina Friedmann, eighty per cent of the Indians of Sibundoyes and Ingas own even now less than five hectares of land. What had happened is that during the last seventy years nearly the entire valley passed into the hands of big capitalists who today manage a milk industry, utilizing ninety-two per cent of the ground, measuring 11,250 hectares. Of these it seems that 22,000 hectares are dedicated to the grazing of some 51,000 head of cattle and 4,000 horses. As for the Sibundoyes and Ingas, they have become famous as botanists, cultivating plants used in medicine that grow (as far as one knows) nowhere else. There are still some 2,500 Indians, speaking Kamsá, living in the districts of Sibundoy and San Francisco, calling themselves Sibundoyes. The Ingas speak Ingua or Quechuo, and they live in the districts of Santiago, San Andrés and Colón. It seems that the Ingas came from Peru. It is further estimated that there are 500 Ingas living deep in the tropical forest and that they are the source of renovation of the medical knowledge of all the other Ingas.

The general acceptance of indigene medicine by the Indians is due to the fact that for every 240 Indians there is usually one Indian doctor who cures without taking money. But there is only one doctor holding an official university degree for 1,500 persons. Meanwhile, for each thousand Indian children born and registered in the district of Sibundoy, 300 babies die before their first year is over. The Sibundoyes and the Ingas have, due to the autonomy which they enjoy in their *resguardos*, become scholars of medicine and botany.

In 1983 the World Wildlife Fund decided to fund WWF Project No. 3093 and to finance as project executant Lynn Bohs of Harvard Botanical Museum, USA. To quote: 'Concentrating first on the most spectacular Sibundoy endemic, specimens were collected of the 12 cultivars of tree datures, Brugmansia aurea and B. sanguinea ... the

Sibundoy Valley is also the supposed centre of origin of the edible crop arracacha (Arracacia xanthorrhiza-Umbelliferae) ... It is hoped further investigation will serve to locate all 14 of the horticultural forms recognised by the Sibundoy.'

For the Capuchin fathers – who had built roads and the hospital – Sibundoy and its Indians has been a rather sad experience, but they will have been glad to read that in 1980 the Indigene Council (Cabilde) then presided over by the linguist Alberto Juajibioy, obtained five schools that employ the services of bilinguist high school teachers in Kamsá and in Spanish, so the Indian Juajibioy, former professor of the University of Antioquia, actually directs a bilingual school for adults in Sibundoy.

# 11 Two Congresses and One Ordination

Perhaps because it is so difficult to do anything really constructive for Amazonia (constructive not only in terms of the economy of the country but most especially for the conservation of the rain forest), many people find that, once they have come into contact with the Amazon region and its specific problems, they remain attached to it for ever.

The Asociación pro Biologia Tropical initiated by Dr J.M. Idrobo, professor of botany at the National University of Colombia, in Bogotá, held two large congresses in 1969, one in Florencia, Caquetá, and a symposium in Leticia a few days later. About 1,400 people from many different countries participated. 'Mrs F.', a wealthy and highly educated American, announced her intention of building a house in Amazonia, as a personal donation, where scientific researchers could stay as her guests, in order to plan for the protection of region. She wanted to create an international park, to save Amazonia's flora and fauna.

Mrs F. knew about Mike Tsalickis' private zoo near Miami. She talked to J.M. Idrobo after the meetings; other naturalists became interested and it was decided that

Tsalickis' animal-exporting activities should be stopped.

Some time later, before INDERENA had fully developed, Tsalickis somehow obtained a tiger which he exported to the United States. In the USA, his buyer could not or would not pay for it, so the tiger went for auction, and as Mrs F. was in Florida, she bought it. She shipped the animal straight back to Mike Tsalickis in Leticia. Not surprisingly, Tsalickis did not know what to do with it, as he could not take it back to its original habitat and no Indian was willing to return it to the wild. Later on, small parrots or even larger ones might be brought back, or freed to fly back into the jungle, because an epidemic had restricted their export to the United States. However, Mrs F., a very convinced conservationist, agreed to pay Mike Tsalickis regularly for the tiger's upkeep, suggesting that it should be kept in a separate enclosure, not far from his tourist hotel. There the poor beast could live long years, growing fat and lazy. Mrs F. paid a high 'alimentary pension' for it, until at long last it died a natural death. She arranged a sort of state funeral for it, much to the amusement of Leticia's inhabitants, and to the annoyance of the Capuchin fathers, who would rather she had spent the money on the education of the Indians.

Mrs F. said to Tsalickis, when she returned to Leticia, 'Every time you export a valuable, rare animal to Miami, I shall have it shipped straight back to you. You are destroying all the animal life in Leticia.'

And so, although Mrs F.'s plan for the international park in Amazonia did not materialize for the moment, INDERENA eventually decided that all national parks in the region should be only partly open to tourists.

Tsalickis continued to export animals until there were very few left, except deep in the rain forest. He continued also with his other private activities, such as his involvement in the ATA airline, part of which was owned by the Prefecture, and all this despite the loss of his status as honorary US consul, due to the stubborn influence of some of the American naturalists.

Monsignor Marceliano acted as the honorary president of the symposium on tropical biology, although privately none of the scientists, in particular Dr Idrobo, was well disposed towards him, because of his old friendship with Tsalickis. They invited Tsalickis to read a paper during the congress on his breeding of monkeys on his island. Which he did! It is worth noting that Tsalickis had built the first tourist hotel in Leticia in which the congress delegates were staying.

The Indians had disliked bringing Tsalickis animals: they had to climb trees at night to get the sleeping parrots, then push them into sacks, where many of them suffocated. Then, provided they were still alive, the beautiful Guacamayos were confined in dark cages, often without water, until finally, after days or even weeks, they could be exported by plane to Miami.

Formerly, every dealer in the port had a parrot or some other tropical animal to attract customers when business was poor. Now the parrots were gone for ever, except for those hidden in the deep rain forests.

During the symposium, the International Animal Protection Society (WSPA) had sent a representative from Switzerland to remind the South American Congress, in which INDERENA was represented, of the new shipping rules drawn up by the International Air-Traffic Association (IATA) as to the correct shipment of live animals. These had been agreed upon and already signed by most airlines. In short, they laid down that, even for short trips, birds in cartons would no longer be accepted. Quite apart from the health certificates which protected the receivers, the animals must be delivered in special crates or cages, whose materials as well as the precise dimensions were specified. Furthermore, the animals must have air, proper food and drink during their journey. Carpenters had to busy themselves in Iquitos and in Leticia, as soon as INDERENA applied these rules and regulations for shipping livestock to the United States and to Europe. At long last Mrs F. and other naturalists were assured that animals and birds would arrive alive in Miami. For a long time Belgium remained hesitant to accept and apply controls in its airports. Britain led the way, encouraged by the Royal Society for the Prevention of Cruelty to Animals.

All this happened, of course, before the mass smuggling of hard drugs, allowing the mafiosi to influence the banks by piling up giant accounts. Then California began to plant its own marijuana, which earlier had been planted and also destroyed in Colombia, especially in Santa Marta. Whole fields of marijuana and then also of coca-bushes were destroyed, but they always seemed to come up again – next-door! Meanwhile the hungry Indian carried on the tradition of chewing his coca-leaves. Years later Dr Idrobo went up with the police in an aeroplane several times to point out hidden drug laboratories in Amazonia. Eventually he gave up, because it seemed a futile undertaking. There were always new laboratories.

Following the congress and the symposium, Mrs F., with scientists from the USA and Dr Idrobo for Colombia, tried all they could to stop, or at least to reduce, the export of animals from the Amazon. Then the botanists began to enquire about the export of the mysterious wealth of plants growing in the Amazon rain forest, whose medicinal secrets only the Indians knew. Dr J.M. Idrobo, in his capacity as Professor of Botany at the National University of Colombia represented the Institute of Natural Sciences at the congress, and its famous herbarium dating from Mutis and his botanical expedition to Mariquita in the eighteenth century. He went to Sutatenza and offered their students a free consultation at the herbarium under his guidance, hoping that a few of the Indian leaders might obtain scholarships for studying botany in Bogatá. He feared, with good reason, that it was too late for the tropical animals, because by then there were few left to worry about. He would protect the plants of Amazonia and also try to protect and help the Indian students. He went to Paris to visit UNESCO and was offered a post in the division for the protection of nature, but he felt that he and his knowledge belonged to Colombia.

The growing network of conservation societies began to enthral the whole world. Agreements were drawn up between people, societies and nations about what *not* to do. They were lucky to find royalty interested. In London Prince Philip and in the Netherlands Prince Bernhard gave their names and their protection to the conservationists, as Queen Victoria had given hers to the Royal Society for the Prevention of Cruelty to Animals (RSPCA).

Dr Luc Hoffmann from Hoffmann Laroche and other international personalities, together with UNO in Geneva, founded the International Union for the Conservation of Nature and Natural Resources (IUCN). This was first based in Morges (now in Gland), Switzerland, and was to act as scientific adviser to the United Nations. It soon acquired the collaboration of some of the most notable scientists in the world. But they needed money for their plans, so the World Wildlife Fund (now the World Wide Fund for Nature) was set up, with offices in London, Switzerland and Washington, among other places. The original function of the fund was to collect money for the protection of wildlife, namely wild animals. The extraordinary spirit of the leading minds – and pocketbooks – of some of those working for the two world organizations made the union a tremendous, lasting success. They decided to publish a list of all the species of wild animals threatened with imminent extinction. It was called *The Red Book*. The animals named in it were not to be hunted, killed or exported without a special licence to be obtained from IUCN. Eventually the same could be done for plant species.

From the beginning, the WWF supported the idea of internationally recognized national parks and financed many such projects throughout the world, including South America.

At that time, the Capuchin fathers in Amazonia, aware that the Indians did not enjoy such official protection, asked if they could at least receive their land in the old pre-colonial reserves. Especially in La Chorrera, they should receive full title to their reserves, the site of their most violent and prolonged suffering. The Capuchins continued to fight for the rights of the Indians.

Meanwhile, the Capuchin fathers in Leticia and in the missions lived very modestly, apart from the Prefect, Monsignor Marceliano, whose way of life, with his residence, garden and secretary-driven car, was a little less modest. Marceliano maintained his interest in the Prefecture's airline, which employed George Tsalickis as pilot whenever Monsignor wished to visit his mission stations. On the whole, the Prelate had reason to be well satisfied with progress in his Prefecture.

In November 1975 Monsignor Alfonso Uribe ordained the Catalan friar Juan Antonio Font as a Capuchin priest. That same year Monsignor Marceliano Canyes signed official papers that handed over all education in Leticia, including the management of the college, to the Colombian government. They were integrated in 1976, at a meeting of managers and district commissioners to approve the triennial plan for all Colombia's National Territories. This meant that they would receive their regular salary as teachers.

On 26 March 1977 an aeroplane loaded with combustible material caught fire on landing in Leticia and, in full view of the horrified spectators, was destroyed along with its crew.

Two months later, the twenty-fifth anniversary of the Apostolic Prefecture was celebrated in Leticia in the presence of Cardinal Arrinal Muñoz and the papal nuncio, Eduardo Somalo. During the festivities in honour of Monsignor Canyes, a number of fundamental difficulties in the Capuchin fathers' work came to light. However, an ambitious programme of activities was arranged to mark the anniversary, including visits by land, river and air to the mission schools on the Lagos, Nazaret and Puerto Nariño.

The jungle, primitive dwelling-place of the Amazon Indians, witness to the birth and death of their forefathers, and hiding-place of a multitude of grave-treasures, had ceased to be a pagan ground. Amazonia appeared now a land of promise. The dispersal of the tribes and races who lived there had but strengthened their attachment to their land, a privileged paradise for those who lived there.

At one of the celebratory events Cardinal Nuñez said: 'If you do not believe in my words, believe in their work. I want the Capuchin fathers to know that there is one person who understands them, is gracious to them and admires their dedication very much. I ask God to give them their reward in Heaven. I feel that the Church has been established here in the Amazon at the price of great sacrifice. I see the work that you, the catechists, embody. You are the crowning achievement of these twenty-five years, the greatest consolation for your beloved priests. At this moment – and I am talking to you with an open heart – you are for me a vision of Christ, which is the bread of life for bishops like me. Father, I give you praise; You have concealed the mystery of Your Kingdom from the wise and revealed it to the humble.'

One of the three Indian leaders, Chief Federico, catechist of Arara, spoke at an official occasion: 'Yesterday, the Señor Cardinal said: "We have not come to bring solutions." Nor do we ask him to bring solutions to our problems. We merely wish to voice them and to offer our comments.'

'We have a profound desire to work and to fulfil our duties, and to show that we are responsible people. However, whenever we come to ask the authorities in Leticia for support, they pay no attention to us. Our families are hungry. The payments we receive as *Curaça** are insufficient. We are therefore asking to be paid enough to feed our families.' (This leader had received free return passage to Sutatenza and had been able to take two full courses, both as free scholarships.)

Another Indian chief, *Curaça* Ricardo, wanted to demonstrate that their concerns went beyond material problems: 'I want to know,' he said in a loud voice, 'if an exchange could be organized. One of our group would go somewhere to see how people worked there, and one of them could come and observe us, to see our customs and evaluate our traditions.'

The *Curaça* of Azulai was considered by his companions to be another St Paul, such were his faith and his oratorical skills. He said, 'The professors of INDERENA do not allow us to cut wood, to hunt or to fish, and they do not let us work on our own farms. But these are the only ways to live in the Amazon. Now another farm has been taken and made into a national park, and INDERENA wants us to leave it. We have a school and we have teachers, but we have no work.... INDERENA forbids us to clear the forests, so we have no wood with which to make products to sell, and we must buy those products from Brazil or Peru. Not only do we have these problems with INDERENA, but now we also have to deal with "Pro-Familia" which is trying to persuade our Indian women to have themselves sterilized, to reduce the size of our families.... This is something which is attacking not only the conscience of all Christians, but also the sovereignty of three countries,

* Chiefs of the Witotoes are called *Caciques*, and *Curaças* with the Tukuna tribes.

Brazil, Peru and Colombia, all of which would see their population reduced by the destruction of the aboriginal cultures, the only native and pure races of Colombia. Of these campaigns, supported by foreign powers, today we Colombians are the ones who are exploiting and murdering our countrymen. I hope that these plans to control the size of Indian families will not bring worse consequences for the peace of the three Amazon nations which have hitherto lived together in friendship.'

This Indian chief was referring to a recent announcement made throughout Amazonia which read: 'Sterilization operations will be carried out on the 6th, 7th and 8th of June 1977, by specialist surgeons from Pro-Familia, and will be entirely free of charge to patients.'

What a shock for the fathers! Within twenty-four hours the bishops and prefects of the ecclesiastical province of Bogotá, united together in Leticia at that time, had formulated a joint reply to this announcement which was circulated throughout Colombia and broadcast by the BBC all over the world. They informed the faithful, and all persons of good will, that:

> The Catholic Church, following Christ's word, has always defended the sanctity of the human being, in particular of the poor, with whom she identifies most closely.
>
> The Holy Father, Pope Paul VIth, in his message for the Day of Peace on 1st of January this year, gave to the Church and to the World the profoundly Christian instruction: 'If you desire peace, defend life.'
>
> Unfortunately coinciding with our presence in Leticia, Pro-Familia's anti-Christian campaign to induce Indian women to be sterilized does not take into account the conscience of these tribal people and their Catholic principles.
>
> We therefore protest in the strongest terms against this abuse, which is, from all points of view, immoral, and an

attempt on the dignity of a human being, whatever his station in life.

Apart from this tactless interference, the silver anniversary of the Apostolic Prefecture was celebrated in an impressive fashion. The Indian *caciques* and *curaças* of the different tribes even had an opportunity to speak up against INDERENA. But one of them, a leader trained in Sutatenza, stood up during the meeting and said, 'We all agree that Peru decimates its Indians; Brazil puts them into a national park; but Colombia educates them.'

The Colombian and foreign conservationists, who were behind the recent efforts of INDERENA, had meanwhile found out that the Prefecture of Leticia owned the Cessna plane and its special flight licence. The Peruvian representative of the WWF, a man of high principle and considerable financial means, had discovered this as a result of tracing the illicit traffic in vicuna and cayman skins from Peru and Colombia. WWF representatives had reported the smuggling of these skins across the Amazon river at night, so that they could be legally exported from Colombia, instead of from Peru, where such trade was already forbidden.

At that time INDERENA had no way of preventing Mike Tsalickis from exporting skins and live animals from Leticia to the United States. Legislation was being prepared in Bogotá, Cali and Medellín, but rather late in the day, and meanwhile the Colombian part of Amazonia was steadily being impoverished. The Capuchin fathers concentrated on saving souls and were following their bishop in his duty to civilize the Indians, according to the Concordat. No matter how they felt about the plight of the exported animals, their vocation was to evangelize the Indians. To educate them was their task.

On the whole, Monsignor Marceliano and his friars were left on their own to deal with the problem of these

Indians from another age. The contrast between the Indians' standard of living and that of the very wealthy families in Colombia was out of all proportion: some families in Bogotá, Cali and Medellín took their money out of the country, often to Miami, and could afford all the luxuries money could buy; the Indians had not even the wherewithal to send their children to school. Only when drunk or dazed with coca did they not feel the terrible difference between their way of life and that of the 'superior' white people.

While the missionaries were striving for the Indians' education, the anthropologists wanted them left alone, so that they could study and write about them. They were, on the whole, opposed to the missions and to the – as they called it – 'deculturization' of indigenous tribes which they saw developing as a result of the missionaries' work. And so, in the Declaration of Barbados (much discussed) in 1971, they even proposed the total expulsion of all missionaries.

The rain forest of the Amazon, ancient habitat of the Indians, was being cleared ruthlessly and the humus even more destroyed, in spite of the various agreements betwen the countries concerned. Overfishing and pollution meant that fish were becoming scarcer. It was feared that the Indians would have no more forests in which to hunt, or rivers to fish, no land fertile enough to grow yucca. They would become easy prey to the Mafia gangs who bought first their coca-leaves and later their cocaine.

The Colombian government paid for some basic education in the mission schools for the Indian children, but the anthropologists fought against the *internados* and also against other mission schools. As a result, the Indians were increasingly deprived of their earthly and spiritual rights.

The Catalan Capuchin fathers stayed and maintained the missions, continuing to teach and to evangelize, as long as their lives were spared ...

# 12   Differing Viewpoints

Unlike the Amazon, where the poor topsoil makes a lasting agriculture very difficult, the Putumayo, or at least the greater part of it, is extremely fertile.

Missionaries visited the Putumayo shortly after the turn of the twentieth century 'It is my desire,' wrote Father Jacinto Mario de Quitó in 1905, 'that on the shores of this river a mission house shall be founded.' A mission was not established there until 1955, when Father Javier de Barcelona came from La Chorrera to select a suitable place for the mission and an indigene *internado* on the shores of the Caraparaná river. He selected a site about ten kilometres from the Putumayo river, and there the mission of San Rafael was established. The following year Father Javier performed the mission's first baptism. By 1959 there were twenty Indian boys and eighteen Indian girls enrolled in that school. Also, the track between the Caraparaná and the Putumayo was completed, and the little community continued to grow there, the fertility of the area attracting a great number of people of many different cultures.

In 1958 Monsignor Marceliano came to visit the mission, accompanied by two members of the congregation of

Mother Laura, Sisters Domus Aurea and Reinalda de la Cruz, who were to run the school. In 1965 a steamship, *Ciudad de Neiva*, landed there with a large load of building-materials. The same year the road linking El Encanto with San Rafael was completed.

In October 1969 Father Javier left and was replaced by Father Miguel Junyet. During Father Miguel's first year, he succeeded in matriculating all his forty-eight pupils, with the very capable support of the teacher, Elba Sajona.

Father Miguel, a Catalan of great persistence and energy, was convinced of the need to encourage agriculture and cattle-rearing among the Indians, and so he set up a farm with 250 hectares of pasture and more than 200 head of cattle, paid for partly with his own family money. After this the teachers and pupils of the school were all provided with milk, cheese, butter, eggs and occasionally good meat. The farm also sold the produce, becoming self-supporting. Father Miguel had good reason to be proud of his achievement. His aim was to persuade the Indians to use the money they earned to breed cattle, rather than squander it in drinking. They were taught how money can be usefully invested.

However, in San Rafael evil forces were at work. Unknown to Father Miguel, the Mafia told the Indians that they had no need to work so hard or to study, and how much easier life could be for them if they just planted more coca. Then, instead of being mere farmers, they would be well paid. Later they would make even more money by producing cocaine for the United States.

One day Father Miguel came back from a visit to Leticia to find that some of his Indians had disposed of a few of their cattle and had not planted vegetables, as he had taught them, nor yucca, as they usually did. There seemed to be more plantations of coca bushes than before. He hoped it would be a short-lived experiment, and he discussed it with the Indians.

In Bogotá, a group of young anthropologists, greatly influenced by pervading Communism, set themselves up in opposition to the missions. Thomas Calle of the National University of Bogotá came to San Jose with his British wife, with the intention of teaching the Amazon Indians to be more independent of the missionaries. They expounded their views on the value of Indian traditions and culture, and caused considerable confusion among the Indians. But they did not stay long. The Indians expected gifts of money or goods from them, according to their custom, but the Calles had nothing to give, so finally the Indians asked them to leave.

The Calles eventually published reports detrimental to the work of the Capuchins, when they returned to London. These were later picked up by a German magazine which carried a rather distorted article, entitled, 'Capuchin monks, the greatest landowners in the Colombian Andes'. Publicity against the Capuchin Catalan father filtered through to Colombia and finally even to the Indians. It only added to the Indians' confusion and sapped their enthusiasm for farm work, which had been most successful under Father Miguel's leadership. So now the Indians began to plant more coca, which they knew could be easily sold.

Despite these various difficulties, in 1979 the mission station of San Rafael became a parish, with a ceremony which included the unveiling of a plaque in the new parish church, to honour its founder, Father Javier de Barcelona.

In 1975 the Indians had participated with great enthusiasm in the second 'Indigene Olympics of the Amazon'. It was even rumoured that a team of Witotoes might be sent to represent their tribes in Colombia at the International Olympics. All in all, there was a good deal of publicity for the third *Olympiades Indigenas* in 1979, which were broadcast widely on Colombian radio and television.

There was another scandal on the horizon. This time it was France which was to see Colombia in an unfavourable light.

A Colombian trader, Julian Gil, and his native companion, Alberto Miraña, were reported missing in La Pedrera on 12 March 1969. It was first thought that they had been killed – perhaps devoured – by an unknown tribe, possibly Yuris, in the area of the Bernardo ravine. Gil was an animal- and skin-trader, one of those whose activities would have ruined the country, and particularly Amazonia, had they not been stopped by INDERENA. He worked with his older brother, and it was this brother who alerted the authorities of his disappearance and led the search for him. Towards the end of April 1969, an expedition set out to hunt for the two men, but it found nothing.

The expedition returned after a violent confrontation with a group of Indians, bringing with it two adult Indians and three small children. They were all completely naked and did not understand one word of Spanish. International news media became very interested in this 'new tribe' found deep in the jungle, speaking an unknown language, which seemed quite unlike any other Indian language. The Indians themselves were first put under military protection in the garrison at La Pedrera, but they fell ill almost immediately, running high fevers, and the Capuchin missionaries of La Pedrera were asked to take them in for medical attention.

Father Romualdo, then director of the mission, accepted them under certain conditions. He and Friar Juan A. Font, tried speaking all the Indian dialects they knew to the Indians, but they did not understand any of them. The fathers were afraid that the Indians would not survive, so they asked Monsignor Marceliano for instructions. He contacted Bogotá, and the Colombian Society of Anthropologists sent an expert, Professor Joaquim Molano

Campuzano, founder of the Jorge Tadeo Lozano University in Bogotá, to see the Prefect, Monsignor Marceliano, in Leticia, from where he travelled on to La Pedrera, to see the famous 'prehistoric' Indians.

Following this, the *Américanistes* in France heard about the 'lost tribe'. The sensational *Paris-Soir* magazine sent Yves-Guy Bergès, one of their best-known reporters, to La Pedrera, to live, eat and sleep with the Indians, trying to communicate with them and win them over with gestures rather than incomprehensible words. The Indians were still very ill, but their high fevers were receding, thanks to the good care given them by the Capuchin fathers.

Bergès flew back to Bogotá convinced that the Indians should be returned to their *maloca* in the jungle. In Bogotá he used his influence to obtain a presidential order to the military authorities in La Pedrera to allow him to take the 'prehistoric tribe' back to its original dwelling-place.

When they heard about this, the Capuchin fathers were, of course, against it. They wanted to give these Indians a basic education and to baptize them. Later it was claimed that this had been done in secret, but there was no truth in this. The Capuchin friars would have liked to have taught the Indians not only to consent but to desire earnestly to become Christians, but they did not understand one another. Not one word – not one gesture – established basic comprehension between them. And so, in spite of all the arguments, the tribe was taken back, as soon as they were well again, to the very place from where they had been removed. The sisters of Mother Laura tried to clothe the children and the woman, but the woman threw her clothes off as soon as she was back in the jungle.

The whole incident highlighted a dilemma: if the Amazon Indians' way of life could be preserved, they could remain naked. But if the clearing of the rain forest continued, the likelihood was that, if they survived, the Indians would be maltreated, suffer illnesses or be exhibited like freaks at a fair.

Two very impressive films have been produced recently, partly in Colombia. *The Mission*, a British film, and *Une Muerte Anunciada* (*A Death Foretold*), after a short novel by Garcia Marquez. Indian children and young people acquitted themselves so well in these that one of *The Mission*'s producers wanted to collect enough money to sponsor their secondary education. These natives did not want to return to the primitive tribal lives they had led before, where girls have no freedom to study. (The problem for philanthropists is that, once an Indian graduates from a university, he seldom has any wish to return to his own tribe, to help them in his turn.)

After his trip to the jungle and the return of the 'unknown' tribe to their *maloca*, Bergès took photographs of the Indian mother, dancing naked and graceful as an elf in the forest. He also wrote a book called *La Lune est en Amazonie* (*The Moon is in Amazonia*) in which he spoke quite kindly of the Capuchin mission, with some understanding of the problems it faced. Another book, *Perdido en el Amazonas*, (*Lost in Amazonia*), by the well-known Colombian author Germán Castro Caycedo, featured photographs of the people involved, including the Prefect and Brother Juan A. Font. Both these books sold well, much to the financial benefit of authors and publishers.

The unknown Indians of the 'lost tribe', frightened, ill and suffering as they had been, received not the smallest compensation. As they spoke no known language, neither priests nor scientists could help them. They stayed in La Pedrera at the mission for such a short time that they were not able to hear, nor could they have comprehended, the story of Christ. They were *not* baptized, by force or willingly.

As for the lost trader, Julian Gil, he was never seen or heard of again, and his true fate was never discovered. The strange occurrence strengthened the belief of the missionaries that it was their duty to work for the general education of the Indian population of the Prefecture.

1978 was an auspicious year: the presidential candidate Dr Belisario Betancur paid a visit to Leticia, and the Andean Pact was signed in Brasilia by Brazil, Colombia, Peru, Ecuador, Venezuela, Bolivia, Guayana and Surinam.

In April 1978 the centre for technical training, the San-Juan Bosco Centre, was inaugurated by the Prefect, by the side of a lake about eleven kilometres from Leticia. This centre had been Monsignor Marceliano's own idea, a place where Indian youths would have the opportunity to acquire manual skills, such as bricklaying, carpentry, painting, mechanics and electrical work. The boys would live at the centre, receiving board and lodging, and tuition for two or three years, free of charge and would be awarded a certificate at the end of their studies. Adjoining the centre was a football pitch which could be used for other ball games, and next to that was the jungle. The boys lived, as they still do, in small, separate houses, eating their meals together in a spacious dining-room. The Capuchin fathers from Catalonia manage and supervise a team of lay teachers paid by the Colombian government. Teachers and boys alike seem to enjoy their life at the centre, especially since the pupils receive a small amount of pocket-money from private donations to the mission.

At the time when the centre was first opened, there was no similar place of instruction for Indian girls, but one was later provided in Nazaret. The building there, housing native catechists, which had been swept away by the flooding Amazon, had not yet been rebuilt, because of lack of funds. Sometimes even Nature seemed to be against the missionaries' work for the Indians.

Meanwhile anthropologists in Bogotá continued to criticize the system of *internados* (boarding-schools), which took native children away from their parents for long periods. When the children went home on holiday every year, they no longer wanted to sleep in hammocks and eat nothing but yucca. Priests and lay teachers agreed that it was not right to take children away from their families, although they were usually brought to the *internados* by their parents. But the distances between their *malocas* and the mission schools were so great that there seemed to be no alternative. It was felt that if the Colombian government was not satisfied with the work of the missions, they should immediately build more roads and also more schools. But it was realized that the Capuchin fathers carried out work in Amazonia which would daunt anyone else. Nobody in Colombia could imagine what would become of the Indian children if the missionaries left Colombia, as they intended to do, as soon as the Colombian clergy could take their place. It was then hoped there would be sufficient Indian teachers available, willing and capable of teaching their tribes.

Leticia continued to develop. In 1979 Julio C. Turbay, then President of Colombia, visited the city to inaugurate a new satellite telecommunications centre and television broadcasting station. That year the census recorded 12,190 people in Leticia and the surrounding rural area, and 2,724 in the city itself.

In 1980 the Institute for Intermediate Education (INEM) was established. In the same year, the Amazon Police Force was created. The Capuchin Prefecture of Leticia under Monsignor Marceliano Canyes boasted forty-five mission stations, together educating nearly 3,000 Indian children and young people.

Father Juan Antonio Font, then a teacher at the high school in Leticia, decided to study anthropology. He wanted to gain a deeper understanding of the psychology of the Indians particularly of the three largest tribes: the Tikunas the Witotoes and the Andokes.

# 13   The Tikunas

Many of the missionaries favoured increasing building activity, thinking that it would be easier to take their Christian message to the Indians if they came together and formed more hamlets.

The Tikunas who lived on the shores of Santa Sofia led a very peaceful existence, dedicated mainly to farming and fishing. This they had learned from their Brazilian friends, who again were much influenced by spiritualism, which has always been very powerful in Brazil. They often came together from Colombia and Brazil, particularly for one of the main feasts lasting three to four days, celebrating the *pelazón*, which means 'pulling out hair'. In this puberty or initiation rite a marriageable girl has all her hair pulled out by her mother or another family member. In the original Brazilian rite, the girl is locked up alone and starved for days and nights. She has to endure all sorts of inconveniences, if not actual pain, to be prepared to accept the hard life of an Indian wife and mother, a slave to her husband from the very beginning.

The husband might lie complaining of his pains in his hammock, while his wife, about to give birth, would continue to work and finally go to the river quite alone.

There she would wash her newborn baby with her own hands. Should the baby be ill or not normal in some way, she is at liberty to drown it or kill it in some other way.

Father Font, in his analysis of pelazón, says, 'In these festivals the young men have an opportunity of getting to know the girls so focusing their affection and aggression accumulated during the monotony of daily work in the jungle.' He does not mention that these Indians (especially the Witotoes) frequently bind up their 'private parts', to avoid them developing fully.

Another reason for gathering together was the small school at the confluence of the Arara and the Santa Sofia rapids. Adults as well as children visited the school, to sell their surplus produce or fish and to join in any festivities organized by the teacher.

Father Font recounted how the natives decided to form a hamlet around the home of the Chief Teodore Angarita (a Tikuna *curaça*,) and his family.

Already, at the beginning of 1968, on one of the missionary visits to the rural school to celebrate a religious festival, the Indians told Father Font of their determination to build a hamlet around their chief's home, giving as their reason a desire to live more closely together. The area seemed suitable for clearing and reminded them of good times in harvesting. Whenever the level of the Amazon fell, the exposed land offered, for a very short time, the opportunity to grow fine crops of rice, maize and so on. On the other hand, it meant that they would leave Arara with its good fishing and fertile land. Yet the prospect of living more closely together seemed adequate compensation for moving from the area.

Everybody joined in building the houses. They tried to build one family home each week and would set off together to search for branches, palm leaves and other building-materials. All the houses were built in more or less the same style.

Father Font made it clear to the Tikunas that he himself
had certainly not initiated the new hamlet; he came to give
as much help as he could, because the hamlet would have
great advantages, but it was also exposed to dangers which
the Tikunas did not even suspect.

Previously the houses were more or less isolated from
each other, and the farming plots (*chacras*) were relatively
near the houses. In the new hamlet, the plots would be
farther away, so fruit trees, palms and medicinal and
other useful plants would not be within easy reach. This
was especially important, as the children would often be
left alone for days at a time while their parents were away;
normally they could easily feed themselves from fruit in the
house, when available.

Father Font insisted that the houses should not be
constructed too close to each other but that enough space
be left for fruit trees, latrines and the domestic animals.
Some of the Tikunas listened to his advice, but others, in
their enthusiasm for the new way of living and in their
haste to be closer together, ignored it.

The missionaries also considered that, if the Indians'
plots, their *chacras*, were too far away, they would stop
going to tend them. Before, they had been used to spending
the whole day in their little farmsteads, where they stored
all their tools for milling flour, their basic foodstuff, but it
was feared that the new arrangement would encourage the
Indians to spend rainy days in their houses, rather than
working on their farms.

The missionaries thought of a possible solution: to concentrate more on producing handicrafts, which the
missionaries could help them to sell. A small museum was
opened in Leticia, where the handicrafts were well exhibited and at the same time offered for sale. Slowly demand
began to grow. The Tikunas took their handicrafts down to
Leticia in their canoes, to sell them to the museum. The
profits they made were used to cover the costs of the whole

community of Arara.

After a while the volume of demand increased still further, and the dealers in Leticia asked for more. Production escalated to a point at which the museum could no longer cope with sales. Contracts were established directly between the Indians and various companies in Bogotá: Artesanias Colombia, Almacèn Cacique, La Ruana and many others. Today there is not a house in Arara which has not at least one occupant manufacturing some sort of handicraft.

The Tikuna chiefs, the *curaças*, were traditionally chosen by the tribal council, new candidates presenting themselves each year. They receive a basic salary from the Colombian government. (Sutatenza used the same system, paying the leaders a small salary for holding their radio schools in Leticia, teaching indigenous adults how to read and write Spanish.) However, as time went by, a feeling grew among former chiefs that the *curaça*'s role should be broadened to encompass the liaison with Leticia and the handicraft outlets. The son of one of the chiefs, Victor Angarita, had been active in encouraging the economic development of the hamlet. He had been studying ways of improving sales of handicrafts, and because he was a Tikuna, his work was accepted and supported by the entire community.

The work of the community could no longer be limited to voluntary communal labour (traditionally called the *guayuria*), which involved clearing the forests and constructing new dwellings. Now the whole hamlet must be kept clean, the plots of land belonging to all the houses swept regularly, as must the streets, the track leading to the school, and all the common areas.

An important new factor, which would have a significant effect on the community's way of life and which at the same time offered a great opportunity, was the tourist trade. Tourists were drawn to the hamlet by the new church made to an entirely traditional Tikuna design – and built by the Tikunas.

Monsignor Valencia Cano, Apostolic Vicar of Buenaventura (later killed in an air crash), called the church 'the most beautiful temple I have seen in all my life'. Soon there was a steady stream of tourists, and it became apparent that they would need to be handled very carefully to avoid disruption of the Tikuna community, which was in the throes of reorganization. The most appropriate person to take charge of the phenomenon seemed to be their chief, the *Curaça*, the highest authority in his tribe and recognized as such by the local district government. He would have to acquire the necessary skills. (It may be that this was a genuine missionary effort to protect the new Tikuna community, respecting the sensibilities of the Indians.) The same applied to the buying and selling of handicrafts, carried out between the Tikunas and dealers or tourists without any intermediary. Here also, leaders were needed who would be trained to take responsibility for the Tikunas' trading activities and for the sympathetic development of tourism.

A request was made to the Colombian district government for the chiefs' salary to be increased, to cover the necessary additional work. This proposal was accepted, and from then on the chiefs of Arara received double the pay of other Amazonian chiefs. In fact, the position of chief in Arara became increasingly onerous: not only had the volume of work increased but the problems were becoming more difficult, and the augmented remuneration, although certainly a help, did not fully compensate for the amount of time and effort which the job now demanded.

The missionaries found investigation into the Tikunas' religious life most difficult. Officially, all Tikunas were Catholics in Arara; indeed, the idea of building a church arose quite spontaneously from within the tribe. Ironically this was to be the one factor of most influence in the cultural changes which affected the tribe, as all the tourist groups visiting that part of the Amazon now wished to include the church at Arara in their itinerary.

The Arara church was the manifestation of a new and beautiful beginning in the Tikunas' religious life. They had all contributed to the construction of the church. 'Now that we have built our own houses,' they said, 'we will build a house between us for God.' An architectural student from the National University of Colombia, Victor Zambrano, drew the design on the ground with a stick and gave the Tikunas guidance to begin with; he and Father Juan Font visited the site a few times to advise, and the Tikunas completed the work on their own.

Every afternoon they met in their church to say the rosary or sing together. Once a month, a priest came to celebrate the Eucharist; literate Tikunas took part in reading the homilies or lectures, and everyone brought an offering, whether a single egg, a flower, some yucca, bananas or fish. At the end of the Mass, the chief collected the gifts together and sent them to the Hogar Virgin de la Paz, an orphanage in Leticia, or sold the gifts and gave the proceeds to the orphanage or to some deserving cause in Arara.

The sermons were always delivered through an interpreter, as many of the Tikunas, especially the women, understood very little Spanish. The Sacrament was preceded by some days of religious instruction, carefully prepared in the two languages. Gradually the frequency of the services and their attendance increased. It was then that the idea of a special institute for religious instruction arose in the Prefecture, or rather the idea of instructing native leaders to act as missionaries in their own communities, speaking the tribe's own language. Bilingual Tikunas were urgently needed. The missionaries recalled Sutatenza's efforts. Federico José Huaines had gone to

Bogotá and from there on to the ACPO's institutes in the Andes, where he took a literacy course, to teach adults of his tribe how to read and write.

Father Romualdo de Palma decided to support the work of initiating the radio schools also in Arara. Federico José Huainas first, and then his brother César, with his wife, Cecilia, had gone to Sutatenza, to the two institutes for boys and girls. As soon as they returned from Bogotá to Leticia, they organized individual radio schools, meeting twice a day, with great enthusiasm. They taught the old Indians how to write a few signs and how to sign their names, and instructed them about topics of general interest, such as hygiene, first aid and vegetable farming.

With the development of education via the radio schools, life continued to change in many ways. After three courses in Sutatenza, Federico Huaines was appointed visiting leader of the various communities served by the radio schools (with a salary paid by the Sutatenza management in Bogotá, as long as donations lasted). The Tikunas had, of course, no idea that the German organization Literacy Aid (through the German Embassy) had paid for their stay at Sutatenza, but the Germans did not pay for the air tickets, which depended on private donations, as did the salaries of the Indian teachers and leaders. For a time it all worked well. The great problem of illiteracy and the need for bilingual leaders of Indian origin with connections and willingness to use the new knowledge for the benefit of the entire tribe seemed to be partly solved in that region.

Now SENA also held courses on cattle-rearing and agriculture. Already Arara owned quite a large amount of grazing land and a number of cattle. One negative aspect of life in Arara was that, in the beginning, not all the latrines were properly maintained; in fact, few families kept them in good order. The result was that intestinal infections were common. Ten young Tikunas died in

Arara without any medical diagnosis or assistance. Tuberculosis was also becoming common again. Few families used water-filters as they should, and the water channel had not functioned properly for years. The small community health centre had no medicine, and the one male nurse, as nobody visited Arara to supervise his activities, had practically abandoned his post. The Tikunas remained afraid of going to the hospital in Leticia, because they thought they would die there. They claimed they were badly treated there and given food they did not like.

The health centre at Leticia had an irrigation system installed in Arara. Unfortunately, it worked for only three months before it had to be repaired, but it was a beginning. Economically, Arara was improving constantly: every house in the village had at least one inhabitant who produced some craft work, whether baskets, hammocks, necklaces or other handicrafts. Most of the money earned by the Tikunas came from the sale of their crafts, and very little, now, from the sale of surplus farming or fishing produce. Several members of the community owned canoes or motorboats which they used to go to Leticia, to visit other villages or to reach their plots of land.

Nearly all the houses now had radios, tape-recorders and record-players, and their owners had smart clothes for special occasions. Some of the Indians had opened shops which sold, as well as necessities, Coca-Cola, lemonade and beer. Alcohol was sold in considerable quantities; a drink called Tatusiño Brazileño was readily available and had a strong alcoholic content.

The Tikunas' lives changed drastically with their new-found purchasing-power; now they could buy tools which previously it had taken them years to obtain. It was a positive change, in that they now had the satisfaction of feeling that they were useful to others, but there were certainly negative aspects.

The community had not been taught how to manage money, which they had never before possessed in such quantities. The missionaries had thought that progress would follow in the wake of the handicraft industry, which would be a source of employment during the long rainy seasons, and that the money thus earned would be used to improve the Indians' houses, their food and their health. However, the opposite occurred. Now that the Indians always had money to spend, they would visit Leticia and squander it on ice-cream, sweets, snacks and drinks for the old people and children. They would buy wristwatches, tinned food, new clothes. Never before was there such a variety of drinks in Arara. Since so much time was devoted to the production of crafts, farming and fishing suffered, with the result that tinned food, rice and biscuits took the place of protein-rich fish, the fruit and seeds which the Indians had traditionally eaten.

As they found an easier way of life, their traditional love for the earth also diminished. The young people stayed at home to paint while the old people went fishing. No longer did they make their earthenware cooking-pots or their small reed baskets: it was easier to buy aluminium pans and plastic boxes.

Similarly the significance of traditional elements of their way of life diminished. The *yanchamos*, paintings made with plant dye onto the bark of the *tururi* tree, which used to provide mats for them to sleep on, had changed into wall decorations. Designs portraying the spirit of Mother Mountain, signified by the shape of a wheel, were supplanted by pictures of jungle animals on pieces of bark, and were snapped up by the tourists. Hammocks, which used to be strictly functional, were now made with less fabric and more *cumare*, a natural plant cord, just to save time and material. In this way the traditional artefacts lost their purpose and originality, if not their quality.

Father Juan A. Font wrote about the Tikunas' cultural

life, about the mythical heroes who had left instructions for their followers to carry out and whose identities were beginning to be confused with those of the Christian stories. In the beginning, the legendary Tikunas were fish, swimming in the brook of Eware Gnupata, whom they called 'Grandmother'. They identified God as the Father of Jesus Christ, and they had a female divinity, Tae, 'Our Mother', whom they had begun to fuse with the Virgin Mary, Mother of Jesus, except that Tae oddly, was identified by them as a man. They gave the title 'Mother' to other beings also, such as Mother Ceiba (a silk-cotton tree), source of good and evil for the magicians. Most animals and plants had their own spirits, which were invoked by the magician, or medicine man, when he entered into a trance, in order to establish which one had the power to heal a particular illness.

The Capuchin fathers saw the danger of trying to convert the Indians to Christianity against a background of dependence and authority reminiscent of colonial days. Rather than wishing to re-create a Christian Church in their own image, the missionaries should try to understand first the Indians' complex psychological and cultural background, and the way in which this was being affected by external forces. A world of insoluble contradictions faced the missionaries: the Indians' traditional subsistence economy, their refusal to save money, the profound effects of a long history of colonization.

The processes of education and development continued in Arara. The two autonomous government organizations, SENA and a body of teachers organized to provide at least a basic education for adults, functioned well and could expand their activities and provide the tools and machinery necessary for the development of this small community. The important thing was that the Indians should understand the need to save money so that funds were available when needed.

Farming was difficult because of the poverty of the soil. Studies were required into the most suitable seeds, and the best products for selling in Leticia. Prices were maintained fairly well, because the Indians preferred to keep their crafts, rather than sell them for a pittance. Unfortunately, as all the chiefs were chosen from the young people, some of their old traditions were in danger of being lost.

A very positive sign for the future was that there were now four active catechists in Arara, who could lead Sunday services in the Tikunas' own language. They organized prayer groups in their homes; attendance at these meetings and at the church services was increasing steadily.

Father Juan Font acknowledged that the missionaries had not understood how to evaluate and make use of the efforts made by the community for the construction of their beautiful church. And so for a certain period the church was not sufficiently maintained, because of the permanent influx of foreign tourists. The Tikunas allowed the building to deteriorate until finally they had to pull it down. It was hard to comprehend how they could do this, when the church had been the origin of the great prestige enjoyed by Arara, and also its primary source of the income from the tourism. However, a grant was given to build and maintain a school where the church had been. It had two teachers and a helper, all bilingual Tikunas; in fact, it was the first school in the Amazonas district to have native teachers who were able to explain the lessons in their pupils' own language. Adult literacy classes continued to be given by the leaders who had been trained or this purpose in Sutatenza. Yet these classes did not really prepare the Indians for the cultural shock received in their dealings with the world of white people.

# 14  Wasted Efforts

The missionaries found that Tikuna Indians seemed to have resisted all efforts to educate them in accordance with white methods. So far, they had shown little interest in studying, striving for academic qualifications or learning any 'white' trade or profession offered to them.

There was a great deal of discussion on all levels about this problem, but the educational programmes were not changed. Perhaps they needed to be amended in order to stimulate more interest, so that at least some of these Indians could be more encouraged to learn certain skills to serve the development of their community. Notably at the San Juan Bosco Centre there were educational opportunities open to the Indians, for their technical training in useful manual skills. This centre also trained Indians from a wide area to become teachers. Unfortunately, these courses were available only part-time. It was thought that they could be extended by means of correspondence courses, adapted to the specific needs of the Indians, and with the great advantage of allowing them to study in their own environment, rather than being educated at some distance. These Indians could thus eventually become the agents of their own development. There was a prime example of this in a group of about seventy Indians who had been following correspondence courses in biblical studies for three years.

Father Antonio Jover publishes a paper called *Acción* which, beautifully illustrated with his own designs, includes the Church calendar, letters with reports and questions from various students, special reports from regional parishes, a humorous cartoon with a moral message, and a section on 'The Word of God', which gives questions and quotations from the Bible for the students to complete. Catechists who finish the course, by sending in correct replies to Father Antonio in Leticia, eventually receive a degree.

This method of evangelizing the Indians is Father Antonio's own enterprise. Guardian of the convent of Leticia, he writes, designs and prints *Acción* without help in a small room in the parish house and convent of the Catalan Capuchin fathers in Leticia. Considering that to become a Capuchin father takes eight years of study and that Father Antonio is the head of the Comisión Pastoral, *Acción*, is certainly an expression of his convinced, fervent missionary activity.

Priests and nuns cannot be totally absorbed by social tasks without depriving the mission of their best capacities. Their daily sacrifice in a hostile climate; the danger of the mafiosi who certainly are against the Indians' being well educated and not planting enough coca; the lack of sufficient religious vocations to help them; and the daily task of precisely recording all expenses, even detailed amounts of costs of daily food, and all the many scholarships that are being financed by the mission: all this has to be registered in the mission's account books. Father Antonio keeps the books and registers the funds received as subsidies for the pupils from the government as well as from private sources and from the community of the order of the Catalan Capuchins.

A difficulty arose when Father Juan Font last returned to Spain. (Every five years the Catalan missionaries travel home, to stay in Barcelona with the community and visit their families.) On the plane he was offered a large amount of money for the mission. He could not accept the money because he knew well that it had not been earned honestly. In spite of all these difficulties, the mission was donating some eighty small scholarships in 1988/89. There were also four university scholarships, all from private donations.

In consideration of so many problems, the missionaries opened a trust-fund, the Centro Pro Indigeno del Amazonas Colombiano (CEPROIAC; Centre for Indians of the Colombian Amazon). And so Father Antonio acquired another bookkeeping task. The Colombian government paid for the basic tuition of Indian boys – for instance, at the St Juan Bosco Centre – but until CEPROIAC, the students had received no money for pencils, paper and other small needs. It was also the first time that it was possible to help *girls*, who received from the fund financial help and encouragement for further studies, especially in Puerto Nariño.

Of the four university students, there was one young man about to finish his medical studies, and a girl training in Bogotá to be a graduate teacher. The Capuchins voted in Barcelona for two Indian girls, who had passed their *bachillerato* in Leticia, to study theology in Bogotá. Previously CEPROIAC had helped them to pass their final degrees at normal school and now it helped and paid their expenses so that they could study.

There are now forty-five adult Indians who, through vacation courses, can act as bilingual school teachers in rural schools. In the San Juan Bosco Centre there are still being offered, free of charge, over three years, scholarships for carpenters, masons, mechanics and electricians. Now there is at long last in Nazaret such a centre for Indian girls, where they receive three-year courses in agriculture, breeding domestic animals, and general information for use as housewives and mothers. The government pays for the tuition, but CEPROIAC gives eighty indigene boys and girls small and larger scholarships to supply pocket-money for notebooks, pencils etc. There is also a plan to open a youth centre in Leticia for peasant Indians to take up secondary studies, as outside the capital only primary schools are available in Amazonia.

After the destruction of the beautiful church built by the Tikunas far fewer tourists came to Arara, so there were also fewer direct sales of the Indians' crafts to the visitors. The Tikunas wanted to sell, but they had no real interest in establishing the best ways of encouraging sales. (This was something that simply did not enter the Indian mind.) Lost income was partly made up by cattle-rearing: under SENA's aegis, several head of cattle were obtained, bringing the total number to fifty. SENA also gave basic instructions and help in improving the living-quarters of the Tikunas.

The Ministry of Agriculture, through INDERENA, helped Arara with a project to plant cedar trees along the roads in the village. There also was a plan to build tanks for fish-breeding. About fifteen villagers were affiliated to a co-operative in Leticia, which sold their products and crafts, gave them half the proceeds and put the other half into a savings account for them. At some time, the Cooperación de Araracuara suggested the plant of Chontadura palms, because the fruit of these palms hold a great deal of protein, produce a very fine oil and can be eaten raw or cooked. But the community of Arara did not wish to accept this idea. Then the community opened its own shop, using community funds, and also received a contribution made by Bienestar Familiar.* There were three other shops owned by individual inhabitants.

Arara now had a 21-kilowatt electrical plant and a channel drawing water from the gorge. The leaders trained by Sutatenza had managed to persuade all the families in the village to obtain their own latrines or septic tanks and to take care of them and not abandon them. The community also owned a motorboat which travelled to Leticia once a week, so that those wishing to sell their products could visit the town and make any necessary purchases. Slowly, improvements were made to help the Tikunas, and their apparently constant problems were gradually being overcome. But, on the whole, the development of Arara, inhabited by Tikunas, as described earlier, was not altogether without problems.

According to the personal opinions of some of the fathers, the Indian is essentially different from the white Colombian. His needs are very modest and he rarely feels any desire to study for a career. First of all, he has to be taught the value of money. But he is the only one who can live on nothing in the rain forest of the Amazon, although now even this is taken from him.

As a child the Indian was not taught to be grateful for what was being done for him in the missions, nor for what the state has done so far for him. He has perhaps little enough to be grateful for. He has not forgotten the stories about Peruvians cutting up the babies of his forefathers as food for their dogs. Why then should he work, with the memory of this tragedy rooted deeply in his soul? And in any case, how could he earn a living now that artificial

---

* Bienestar Familiar was set up in 1969, following a suggestion made by the President's wife, Doña Cecilia de la Fuente Lleras, mainly to protect children and for the welfare of Colombian families and orphans. This organization has successfully intervened in the problem of Colombian children being adopted in Europe and the USA. It also helped children after the terrible eruption in Armero of the Volcano del Ruíz, in November 1985, when some 23,000 people were buried alive.

latex has replaced the tears of the Brazilian Hevea tree?

In that part of Colombian Amazonia the only industry employing native Indians was the tourist trade. Food for the hotels was still flown in to Leticia from Bogotá or brought from Brazil, rather than being grown locally, which would have provided employment as well as a healthier way of life. The Colombian-Dutch Cooperación de Araracuara, set up in 1977 with the financial and technical collaboration of both governments, at first produced very fine vegetables also began to breed giant turtles quite successfully, but management did not seem inclined to contribute towards the education of the Indians by helping the Catalan Capuchin fathers now well settled there, as this was not in the agreement of the participants.

Still, at some distance from the cooperative, a new missionary school was inaugurated for the Indians at Araracuara in 1977 by Father José Maria Clarassó, who, at that time, selected the students to be beneficiaries of the scholarships the German Embassy was offering them at Sutatenza. This Capuchin father saw to it that the Indians got onto the planes and travelled, via Leticia, to Bogotá. As these Indians were poor peasants, they had never been in a plane before and were frightened at first. At that time, the Capuchin missionaries and the people of Sutatenza were getting along very well. Later on, SENA grew more interested and took Father Clarassó's place, as he had been transferred to Nicaragua. But, sadly, there was practically no co-operation between the scientists paid to work for Colombia at the co-operative and the unpaid priests trying to further the education of the Indians.

Monsignor Marceliano Canyes, who came to give his blessing to the new mission school at Araracuara, was – until he handed the responsibility of education over to the Colombian government in 1975 – alone in charge of the welfare and education of some 3,000 pupils at the forty-five mission stations of the Capuchins in Amazonia. The very rapid expansion of this work had also been made possible by the availability of George Tsalickis as the Prefect's pilot for so many years. He was probably the first to fly the length and breadth of the Colombian Amazon.

At about the same time, in December 1975, the Colombian government created reserves for the Indians: the Witotoes of Monochea were given 377,000 hectares, those of Puerto Sábalo los Monos 304,000 hectares, and the Andokes some 370,000 hectares.

Title to La Chorrera, the giant 'hacienda of the Putumayo' of nearly 5 million hectares, was not clear, as the Caja Agraria, the Agrarian Bank of Colombia, claimed some rights through an earlier purchase. The arguments continued.

In June 1977 the Anglo-Colombian scientific expedition began its work researching better ways of using the poor Amazonian top-soil. In Araracuara some useful experiments were carried out, only to find that the Indians (the Andokes) had grown better yucca and much bigger carrots! They also experimented with different types of oil-producing palm-trees and with the cultivation of the 'chocolate' tree. It was decided to concentrate work more on the production of cocoa and to establish which varieties would be best suited to the Caquetá. Meanwhile, it was rumoured that the Dutch were really searching for uranium, but so far without success. This was further consolidated by a series of threatening telephone calls, inviting some of the Dutch initiators to leave Colombia soon or be shot!

The Dutch government gave its approval to concentrating all experiments on cocoa production, and Colombia hoped that cocoa might become as important an export produce as was its coffee. At present Colombia does not produce enough cocoa for its own domestic consumption.

In 1981 the indefatigable Father Juan A. Font held the

first short course for Indian aspirants wishing to become Catholic catechists. It was held in Araracuara. He founded the Franciscan Youth Centre in 1982, called JUFRA in Leticia, which he still directs, in addition to his teaching duties at the high school in Leticia. This same JUFRA is now busy organizing an interchange of young Christians between Catalonia in Spain and Leticia in the Colombian Amazon. There were bazaars to sell artefacts made by JUFRA youths, to help raise funds to enable four of them to travel to France, in December 1988, for a giant congress in Paris. The young people gathered there voted that they considered Indian youth to be entitled to secondary education.

But before this plan materialized, on 21 September 1981, an accident occurred which had a profound and damaging effect on Monsignor Canyes' life and on the progress of his difficult mission work. George Tsalickis was trying out a new light aircraft when, just before landing at Tabatinga, the plane caught fire. His co-pilot was able to eject and survive, but George Tsalickis perished with the plane, which finally plunged into the Amazon.

Monsignor Marceliano had been able to fly from Leticia to La Pedrera in ninety minutes, compared with eighteen days' journey by river. The death of his pilot changed all that. He was at a loss. It was doubtful if the Colombian government would help, and in any case, he felt too tired to fight this battle. He had given nearly fifty years of service to the Church in Colombia and had succeeded, with his priests and with Sutatenza's help, in eradicating illiteracy from the part from Colombia for which he was responsible. The Prefecture held 121,240 square kilometres, and there were about 15,000 Indians living there. Now at least they had all received primary education and could read and write. But Monsignor Marceliano faced a very serious eye operation, as did one of his oldest and best priests, Father Cristobal, who for that reason had retired from La Chorrera. The Prefect was suddenly very conscious of the approach of old age.

This was not the only difficulty: there was a new evil in Colombia. Now the Mafia not only traded in dried-up coca leaves; they had also set up illicit factories all over the region to produce cocaine. Purchasers in the United States seemed ready to buy all South America could produce, and pay well. Soon the evil was deep-rooted and involved many people and even important banks.

Monsignor Marceliano held the funeral for George Tsalickis and had an engraved plaque put in the church to commemorate his service as a pioneer pilot in the Amazon. Tsalickis would not be replaced. His brother Mike attended the funeral service with his family, distraught with grief.

'I shall leave the country as soon as I can,' he said to Monsignor Marceliano. 'All is spoiled for me – now.'

Many of the Indians working at the hotel built by Mike Tsalickis, wanted him to leave. In fact, they threatened to kill him if he did not. They had had enough of his demands on them: they had always resented having to perform their dances for the tourists who stayed at the hotel. They had found it humiliating at a time when the Capuchin fathers from Spain were trying to enhance their self-respect by giving them a white man's education. These Indians had also discovered that the export of monkeys, which Tsalickis still held in captivity, was now officially forbidden by mighty INDERENA. Did the Indians suspect as did the ambassador of Colombia at UNESCO in Paris, that Tsalickis' was exporting cocaine with the animals?

But what could be done? Mike Tsalickis had a Colombian wife and six children, all born in Leticia. He also employed some hundred traders, who would lose their jobs if he lost his freedom. So the authorities kept silent.

Meanwhile, the old Santa Sofia, now called 'Monkey Island', originally bought by Tsalickis in his wife's name as a breeding-place for monkeys to be exported and sold to the United States for laboratory experiments, was leased out for tourist accommodation.

Finally, the Tsalickis family left for Miami, where they had piled up a nice fat bank account. Then it was rumoured that Mike had been killed in Brazil. Still later, they heard that he was, in fact, far more interested in drugs than in wild animals. But in 1987 the news was confirmed and widely published that he had been caught in Miami and was in prison in Florida. Although Tsalickis had argued that he did not own the boat in which he had been caught and that he knew nothing of the giant load of cocaine which it carried, bail was refused. They did not let him go free, no matter how much money his family offered to get him out of prison. At long last, his career was finished.

Another sad story which threatened to affect the morale of the Capuchin fathers concerned Father Miguel and the beautiful farm he had set up on land belonging to the Indians, by the mission school and *internado* near San Rafael. As land in the Putumayo is very fertile, the mission had become self-supporting. But while the father was on home-leave in Catalonia, the Indians working in the reservation had planted more and more coca and less yucca. Their main foodstuff, manioc was made from the yucca roots; vegetables had been planted; they had had meat, milk, butter and cheese, and were healthy. But farming and agriculture did not seem to interest them any more. They had actually slaughtered or sold the cattle reared so painstakingly with Father Miguel's financial and moral support, and they would also have sold their land had they been able to do so legally. Fortunately, the land is part of the Indian reservation, meaning it cannot be sold to outsiders but passes from father to son or to other Indians.

It was true that these Indians had now more money than ever before, but very few of them learned how to invest it wisely. Father Miguel observed sadly that many of them were wearing fine wristwatches.

The government in Bogotá cut the mission's budget, and the Mafia's hold on the country strengthened apace. Laboratories sprang up deep in the middle of the rain forest, and the government convinced some Colombian botanists and anthropologists to fly over the forest in helicopters and indicate to the police certain spots where they suspected laboratories for the distillation of cocaine were hidden. For everybody knew by now that the Mafia processed coca leaves, or coca paste, into cocaine for export by private planes to the USA, especially to Miami. It was also rumoured that they kept private zoos for their entertainment and also their protection, whenever needed.

Suspicion grew that the Mafia was linked with the *guerrilleros*, selling weapons in exchange for cocaine. Nobody knew exactly what the deal was, but it seemed quite pointless to denounce the Indians for planting too much coca and not enough yucca.

It is now necessary to recount what happened very recently to a Tikuna tribe to the north of Brazil in the Amazonas region. There is a fine settlement of Tikunas near Leticia, at a distance of six kilometres on the road of Tarapaca, with their own school and church and good houses or (*malocas*). They were friends or relations of the Brazilian Tikunas, who came to visit them frequently. In April 1988 at least twelve of these Tikunas, including their children, were killed, and thirty more heavily wounded, while peacefully working in the fields. Suddenly, white settlers, some twenty men, heavily armed, under the personal leadership of a white latifundist, an owner of much land and dealer in woods, shot them down. The reason seemed to be that two years previously the Brazilian governments had signed a decree giving to the Tikuna tribe some 103,000 hectares of land, bordering territories situated between Brazil and Colombia or Peru. However, as frequently in the past, this reservation had not been immediately precisely marked by competent authorities, and so permanent difficulties arose between some 17,000 Tikunas living on their reserve and white settlers wishing to enter and occupy it. It would have been the responsibility of the official guardians of the indigenous population, the FUNAI, to claim a precise demarcation of the reserve, and to take more care of the hygienic installations recommended by medical authorities of the World Bank.

FUNAI, the organization charged to guard Indian interests in Brazil, demanded that the assault should be punished as 'genocide'. The Brazilian police caught eighteen of the aggressors with their leaders. All said they had shot the Indians in self-defence. No single example is known in Brazil in which Indians have won a case against white settlers.

There is a widespread rumour that gold may be found in this region, and this possibility has always attracted people of all kinds, some ready to commit any crime just to obtain gold. Such prospectors often bring epidemics of influenza, measles and other diseases against which the Indians have no natural defence or adequate medicines.

Medical experts from the World Bank had previously found that FUNAI had forgotten to take care of sanitary and medical care in the reserve. Now the suspicion exists that FUNAI may have known there was a danger the Indians might be violently attacked by white settlers.

Acting as the Brazilian representative of all Tikuna interests, Bishop Erwin Krautler, president of the Church's Indian Mission Council (CIMI) had courageously published a note demanding that the police should investigate the precise role of FUNAI during the recent assault on the Tikuna village. The bishop considers it quite possible the FUNAI is also at fault. He has many times endeavoured to stop white settlers from invading Indian reserves.

# 15 Wicked Deeds and Other Complications

No effort is in vain for those whose lives are committed to their religion. Anybody who has not experienced or observed the demands of a strong religious vocation would find it difficult to understand Mother Laura Montaya's words: 'Destroy me, Lord, and over my ruins build a monument to your glory.' Madre Laura, who founded a missionary order of sisters to serve the South American Indians, felt strongly that all her suffering was for the glory of God, and she hoped to bring to God the souls of the Indians she loved. And Bergès wrote about one of the Capuchin fathers he had met in Colombia who had renounced his large inheritance in Spain in order to serve the poorest, the Indians, by becoming their missionary.

There seemed little improvement in the daily life of the fathers. Suffering and danger were becoming ever-present parts of their lives. So it was little known that upon the hit-list of the *guerrilleros* was the name of Monsignor Marceliano. One of the fathers was captured and imprisoned by the Mafia and for three days was forced to dig his own grave, to persuade him to make sure that the Prelate would personally attend a certain meeting.

Early in 1984 Father Antonio was waiting for his pupils to join him in Araracuara, where he was running a holiday course as part of the correspondence studies in their religious education. During the course, much of the time his Indian pupils were so petrified with fear that they were unable to reply to his questions. They would sit, clutching their rifles, night and day, ready at any moment to defend the mission from a guerrilla attack.

At that time, left-wing *guerrilleros* were protecting what was probably the largest cocaine-refining plant in Colombia, in a place called Tranquilandia on the Yari river. In spring 1984 the place was raided by the police on the order of the Justice Minister, Rodrigo Lara Bonilla, and sanctioned by the President Betancur. The plant's location had been revealed by satellite.

Viewed from the air, the Amazon basin in Colombia is a lush emerald-green carpet which extends across some 5,000 square kilometres of practically uninhabited land. The picture has been spoiled by brown scars punctuating the jungle carpet, wildcat airstrips for invisible laboratories using private small planes. Now, with the use of satellites, these laboratories can no longer be hidden.

When they raided the plant, the police discovered that the whole village was dedicated to the work of refining cocaine. They found nearly fourteen tons of processed cocaine, marijuana cigarettes, tons of chemicals used in the various processes, and invaluable documentary evidence, including files of receipts, totalling millions of dollars. Everything was in the ownership of a handful of powerful Colombians, operating under the protection of the guerrillas.

In Tranquilandia, which the press called 'Villa Coca', the police discovered vast numbers of cars and tractors, and an entire fleet of helicopters and light aircraft. The village had every modern comfort and convenience: washing-machines, televisions, microwave ovens to dry the coca leaves, and depots full of electrical goods used in the refining processes. Most impressive was the discovery of fifty guerrilla uniforms, seeming to confirm the existence of an alliance between the Communist guerrillas and the mafiosi. Documents they found showed that the centre was owned by two well-known and fabulously rich members of the Mafia: Escobar and Ochoa. Needless to say, the bosses were no longer in Tranquilandia, having left a short time previously in a specially equipped jet plane.

Not long after the raid, Lara Bonilla, Minister of Justice, was shot dead from a passing car in Bogotá. Further proof of the link between terrorism and drug-trafficking came a few days later, when one hundred armed guerrillas, belonging to the notorious M-19 group, carried out a raid on the city of Florencia. They held the city hostage, with its 500,000 inhabitants, for several hours until driven away by the Army.

Florencia was founded by the Capuchin fathers, guided by their conviction to serve Colombia. Now, behind the tropical façade of the jungle are barren areas to which vegetation can never return, areas of concern to ecologists throughout the world. Here abandoned coca fields have been sprayed, just as the marijuana fields were once sprayed in Santa Marta.

An article in *Time* magazine in February 1985 claimed that ninety per cent of the world's coca was grown (and processed into raw coca paste) in Peru and Bolivia, only ten per cent being grown in Colombia, where the paste was compressed into cocaine and then smuggled into the United States. It was a very lucrative business. Laboratories sprang up all over the world, even as far away as Switzerland, near the Roman Catholic Fribourg.

In Amazonia, the tragedy of the farm organized by Father Miguel in the Putumayo was not an isolated case. Thousands of hectares of forest, often precious rain forest, were hastily cleared and planted with coca shrubs. Once they had been cleared, the Indians never return to the same plots, so these gradually turn an ugly yellowish-brown colour. When the coca bushes were green, they were almost impossible to detect from the air: even if they were detected, there was no non-poisonous spraying agent available to kill them. Botanists and other scientists handed their lists, which they had received from scientific bodies, to the police, telling them which spraying agents were dangerous to be used.

According to the Bogotá newspaper *Espectador* (2 December 1986), J.M. Idrobo, then Botany Professor at the National University in Bogotá, had recently discovered a larva (grub) called *narcomején* Eloria Noyesi, that eats coca leaves. He discovered it in the coca fields at San José del Guaviere and suspected that, as Bolivia and Peru had always produced better, stronger coca, rival interests might be jealous of, and attempt sabotage in, the increasingly large coca fields in Colombia and Bolivia, which was supposed to maintain its hidden laboratories and work the paste received from Peru. Eloria Noyesi's capacity seems astonishing: although quite small, it can consume vast quantities of coca leaves; it eats nothing else; and botanists know so far of nothing else capable of destroying entire fields of coca in a short time without any danger to man or beast.

The difference in attitudes towards coca is one of the problems so far unsolved, affecting relationships between Indians and white men. The Indians have always regarded coca as their friend; even now they need it for food. White men, in their limitless greed for money, had discovered how to extract cocaine from the leaves to such an extent that cocaine was now said to represent twenty-five per cent of Colombia's exports, officially coffee, cut flowers and emeralds.

Coca leaves nourish the Indians and help them combat

fatigue. During colonial times, and even afterwards, they would be paid in dry coca leaves. As a food it was excellent, providing more protein than any other fruit or vegetable which formed their diet, and useful vitamins. The alcaloid substances contained in the leaves were eliminated by the body. Because of its usefulness to the Indians, for whom he has always cared, J.M. Idrobo intends to carry out further research into the life-cycle of Eloria Noyesi, in order to be able to teach them how to combat the growing of coca without using harmful chemicals. He believes that the Indians know many mysteries of nature in the Amazon region, so far still undiscovered by white men.

Drug-addiction remains the greatest worry of scientists and priests alike; it is growing steadily in Colombia too. In former times the drugs were all exported as a means of earning more money. Now the Indians are beginning to smoke a cheap and very dangerous substance called *Bazooka*, a bi-product of the cocaine-manufacturing process, mixed with various chemicals. Physicians and other professionals think that, if there should be a general liberalization of cocaine, the chemical industry might produce even worse substances, such as 'crack'; others feel that, if prices for drugs were to drop sharply, interest in the drug business might cease.

When Father Antonio returned to Leticia after teaching in Araracuara, he hoped that his seventy Indian pupils would continue their studies of the Bible as catechists, but at times he and the other missionaries despaired.

Perhaps the Capuchin fathers decided to pay for two Indian girls to be educated further and to continue studying theology in Bogotá because they hoped they might fight against the spread of the cocaine trade in their tribe. The fathers may have read in 1984 that the Colombian Mafia had made an offer to the conservative government to bring back some $3 billion profits from their drug-trafficking activities – into Colombia in exchange for a complete amnesty and cordial treatment by political and ambassadorial circles in the country. Lopez Michelsen, President of Colombia between 1974 and 1978, was said by *El Tiempo* to have proposed the Betancur government to accept the Mafia's offer, saying: 'If these people want to give up their laboratories and coca farms, I believe we should accept the lesser of two evils, and do business with them.' He felt this would be the only way to reduce drug-trafficking in Colombia.

It is doubtful if the Mafia would have dared to make the offer, had they not been backed by some, at least, of the international banks.

President Betancur, however, refused to negotiate, a refusal which led to the assassination of the valiant Rodrigo Lara Bonilla, Minister of Justice. And, in recent times, many of the 'cartel' in Medellín have been killed.

# 16 Martyrdom

Unlike Colombia, which has had a special autonomous division of the government responsible for Indian affairs for many years, Ecuador has only recently created an Indian Affairs Office. This came about following a visit by the Pope, who demanded greater concern for the native population. Although Ecuador did sign, in Geneva, the International Convention on Tribal Peoples, this is practically never invoked.

This brings me to my reason for writing this book. Where nobody is willing to take action, missionaries, friars and in particularly Franciscan Capuchin fathers volunteer to help. They were asked to establish a mission in the Shushufindi Valley, where oil wells were being drilled, a mission which the fathers completed in 1977. According to *International Survival*, the Indians in that area harboured a grievance against a company called Palmeras del Ecuador SA, which was extending its plantation of African palms over 10,000 hectares, quite near the oil wells. Since the growth of the palm-tree plantation and the expansion of the oil wells, over the past ten years, almost all the game in the area had been driven away, and many of the fish in the rivers had been killed by pollution. The economies of both the Siona and the Secoy tribes, numbering about 300 persons, had been devastated.

Bishop Labaka, a Spanish Basque of the Capuchin order, was aware of the opposition of some small tribes, many of whose members were still illiterate, to the cultivation of African palms west of the Andes. The Indians argued that these palms had been grown successfully for many years on the coast. Now the Indians were suffering from pollution and from skin diseases which they attributed to the insecticides sprayed onto the palms. Eventually the Ecuador government listened to them but it still refused to withdraw the oil companies' licences. The oil boom in Ecuador had begun some ten years earlier; Texaco, in particular, had built a series of roads, pipelines and even entire towns, devastating the forest and causing continued suffering to the Indians. The agreements signed with the oil companies had not protected these Indians from losing their land.

Bishop Labaka, head of the mission, wanted to pacify the Indians – a dangerous task. Before setting out on his journey, he telephoned his friends in the Capuchin order in Barcelona to tell them of his plans. He also told the Capuchin fathers in Colombia that he had decided to evangelize certain small tribes opposed to the oil companies. He informed them that he was well prepared for the trip. The oil company would provide a helicopter and a pilot, because naturally they had a vested interest in soothing the tribes. Before landing, he intended to throw down photographs of himself and his companion, with messages in Quechuo and some small gifts, hoping that the tribe would understand that he was their friend. He was thinking of taking a nun with him, if he could find someone who was ready to offer her life to convert the tribes to Christianity. He made it clear that he himself was prepared to die, if need be.

Sister Inés Arango Velázquez was a Colombian from a

very devout Catholic family in Medellín. She and her sisters Fabiola and Cecilia all belonged to an order called the Congregation of the Capuchin Tertiaries. She had entered the order in 1954 and become a highschool teacher, but her vocation was to become a missionary.

In 1973 Inés had written to the provincial mother superior asking to be sent as a missionary to help found the mission in Mitú. At that point, the Capuchin fathers of the mission of Aguarico in Ecuador put in an urgent plea for the foundation of a mission in Shushufundi, near the oil wells. Sister Inés was one of the first sisters sent there, but only a few months later she was transferred again, to another mission of Rocafuerte, on the River Napo. Nuns belonging to the Franciscan Missionary Action Group had been working there, before being replaced by the other sisters. From Rocafuerte Inés was able to visit the Amazon forest regularly, to exercise her missionary zeal among the Indians. She was accompanied by Father Alejandro Labaka, and together they brought the word of Christ to the small Auca tribe.

In 1985 Father Alejandro Labaka was nominated to succeed Monsignor Jesus Langarica as Bishop of the Vicariate of Aguarico. He continued his work among the Aucas from his new home in Coca. Sister Inès, after six years as mother superior in Rocafuerte, was also transferred to Coca, to take care of the community there.

In early July 1987 Sister Inès represented the mission of Aguarica at the Third Congress of Latin American Missionaries in Bogotá. She returned to Quito on 17 July. Shortly after her return, she was contacted by Monsignor Labaka, who asked her to fly with him to the land of the Tagaens tribe.

On the first attempt to make the trip, the helicopter had to turn back because of poor visibility. On the morning of 21 July 1987 they tried again. The pilot was able to set the helicopter down safely in the Tagaens' territory, and the two missionaries alighted. Monsignor Labaka had told the pilot to take off when he raised his right arm, but to come back within ten minutes. He gave the agreed signal, and the helicopter took off. After ten minutes the pilot tried to return, but he had lost his bearings and was unable to find the place again. He radioed the mission and the oil company for advice, and it was decided that he should return to base and try again the following day.

On 22 July the pilot and a Capuchin friar from the mission, Father Roque Grandez, flew out to reconnoitre. A terrible sight met their eyes. Monsignor Alejandro and Sister Inès were lying lifeless on the ground. The helicopter returned immediately to alert the mission and the oil company and to enlist the help of the Army. Two Army helicopters, with twenty soldiers, went back, with a third helicopter carrying Father Roque Grandez and Father José Miguel Goldarez from the mission.

The bodies were taken away with great difficulty. Monsignor Labaka had suffered fifteen spear wounds, Sister Inés three.

Medellín in Colombia and Barcelona in Spain were notified; the funeral was attended by the mother superior, Inès's two sisters, a large number of Indians and many other groups of people who knew and respected the two martyrs. They were buried side by side in the cathedral of Augarico.

In an article in the *Revista de Misiones*, the Capuchin Tertiaries gave thanks to God that one of their order had been permitted to become the first Colombian martyr.

# 17   What Will Be the Future of the Amazon Indians?

What more can be done to help the Amazon Indians to develop their talents and abilities and give them a better standard of living, through higher earnings? Peru has recognized Quechua as its second language, but Colombia's Indians speak too many languages for a similar recognition. However, Colombia has trained bilingual leaders, who work in the schools teaching their own tribes. All South American countries have recognized *reservados*, land belonging to the Indians, but they are slow in determining frontiers and rights. As soon as gold, petrol and perhaps even uranium or diamonds are discovered or suspected, the Indians' land is vulnerable to exploitation. How can they defend themselves if their rights are not well established?

Colombia has shown the treasures of its unique Museo del Oro in exhibitions world-wide, giving opportunities for foreign countries to admire its indigenous arts and skills. In this regard, the Capuchin fathers intend to organize, together with the Banco de la Republica, an important exhibition in Seville, Spain, of Indian crafts from different tribes of the Amazon for the centenary celebration in 1992, celebrating 500 years since the discovery of South America by the Spaniards.

Furthermore, INCORA has been distributing generous lands to the Indian tribes, and it is hoped that funds will be granted for the education of the Amazon Indians. When President Betancur invited the Spanish Prime Minister, Gonzales, they both found that, of all Colombia, there was the least analphabetism in Amazonia. This is what the misionaries have achieved in the most difficult region of the changing Amazon river, as well as in the territory of the Caquetá and the Putumayo streams.

In Colombia's Amazonia the Tikunas are the most numerous tribe, but they are originally rooted in Brazil, while there is no doubt that the Witotoes have their cradle in La Chorrera. The Capuchin Catalan fathers built their first missionary school here, on soil bathed by the blood of the tribes reduced within ten years of rubber exploitation from 50,000 to 5,000 Indians. The Casa Arana has faded out, but the effects of its crimes are difficult to erase from the memories of the existing members of the tribe.

The Witotoes believed that they were the 'chosen tribe' whose sons were dispersed throughout the Amazon. They left their own lands and began a long pilgrimage because of the permanent interference of outsiders, impostors who disturbed their way of life and deceived them with false promises. If this is their prime tribal conviction, the following story, dating back to 1924, can be understood, justifying the hope that the Witotoes will safeguard the paradise which is now their property:

In 1964 the *Caja Agraria* (Agrarian Bank) bought from the then existing Casa Arana, the so-called '*Hacienda* of the Putumayo' a property of over 3½ million hectares, for the sum of $160,000 US. This transaction, when they paid the rest, was therefore old in date, as a down-payment of $40,000 had been made some forty years earlier. Such important transactions take a long time in Amazonia! It was not until twenty years after this, namely in 1984, that Dr Mariano Ospina H., son of a former president of Colombia, then director of the Caja Agraria, took up the matter of the old contract, thinking that something should be done about the very fertile land of the Putumayo, and also about the remaining Witotoes and their different tribes. He noticed that there was still the same lack of roads as before.

Dr Mariano Ospina decided that the Caja Agraria should open up an experimental station where once had been buildings of the Casa Arana, forgetting that the remaining ruins and buildings really belonged to the order of the Catalan Capuchins. They had been officially presented to the order after the battles between Peru and Colombia, and had been accepted, in the name of the order, by Father Estanislao de las Corts. However, in 1951 the waters of the Igaranaparaná river submerged the first floor of the mission school, and as the flooding continued, the *internado* was transferred to the hills on the other side of the river, where palms grew in abundance, in a fine climate and good earth. It has remained there to this very day, much enlarged, with a beautiful church, already described.

When he was director of the Agrarian Bank, Dr M. Ospina intended to make very large donations to the Witoto tribes, but the Witotoes refused. They went back to Bogotá, undertook a great deal of historical research work and decided they wanted to receive their land, their old *reservado*.

The Capuchin fathers, silently observing the building of the Caja Agraria experimental station, had always doubted that it would be possible to get the Witotoes to collaborate and give up their legal rights to their historical reserves. The winning argument of the clever Witotoes was simply that the Casa Arana had never owned the land and therefore was not able to sell something they did not really possess.

Meanwhile, Father Cristobal, close to retirement, had

withdrawn from La Chorrera after a serious eye operation in Bogotá and was considering working in the parish of Leticia. A secular priest and a full-blooded Colombian from Medellín, Father Daniel Restrepo Gonzalez, applied to take over the management of the boarding-school and mission in La Chorrera. He espoused the Indians' cause and advised them how to proceed to obtain their rights. Meanwhile Monsignor Marceliano kept silent, for reasons of health.

The Witotoes refused to agree with the Caja Agraria, and as Dr Ospino had resigned, President Barco charged the Colombian Institute of Agrarian Reform (INCORA) to purchase the land from the Caja Agraria and give it to the Witotoes as their legal *reservado*. So the land reform in Colombia gave to the Indians as their sole property the Resguardia Indigeno Predio Putumayo, which means that the Putumayon *hacienda* of nearly 6 million hectares now belongs to the Witotoes.

INCORA states that there are some 10,000 natives grouped into 2,065 families, who live on their own land and will be the sole proprietors of 5,868,477 hectares. The general manager of INCORA says that these Indians dedicate themselves to agriculture, to the collection of wild fruits and to hunting and fishing; also that the communities of the Witotoes, Murui-Muinane, Ocaina, Andoke, Carijona, Mirana, Tikuna, Cabiyari, Inga, Sionay and Letuama have lived on this land of their reserve for more that one hundred years and that they are always applying their methods of conserving their natural resources.

Research in Colombia discovers more and more that the agriculture developed during centuries in the Amazon rain forest reduced to a minimum the dangers of an ecological hecacombe. The Indians selected their trees to be cut down without removing them immediately, but only as they needed the wood, so the trees stayed there and protected the ground from rain and strong sunshine. These same methods are being observed by Indians now, provided settlers will not disturb them.

In the rain forest of Amazonia, every single plant and each tree fights for light, which filters sparsely through the flat ceiling formed by the leaves of giant trees. From the tree branches hang the air-roots, lianas and bindweeds, which are organs developed through centuries, and take up the beneficial substances in the water before they reach the ground. In this humid rain forest each hectare contains hundreds of different species, but the ground can maintain only a very few herbivorous animals because, lacking light, there is very little or no grass.

But when settlers come, they start cutting trees in a large area, and with them fall the lichens and the orchids which usually support the insects, salamanders, monkeys and parrots, and so their droppings will no longer feed the foliage resting on the ground and protecting it. If settlers come from the Andean region of Colombia and establish themselves spontaneously in regions such as the Caquetá, they will cut down large sections of rain forest, take out the trees and sell them and sow maize, kidney beans and bananas. They then fence the area they have cleared and establish themselves permanently. The first harvest will be splendid; then, as the strong rains wash out the nutritive substances, and the tropical sun dries out and hardens the ground, the result must be the sterility of the further cultures and the loss of their capital, which may force the settlers to sell out to rich commercial dealers. They have not considered the method of the Indians of the Amazonas who are supposed to have lived there for some 11,000 years, hunting, fishing, fruit-collecting and also cultivating this meagre earth for over 3,000 years, without destroying the ground and the rain forest.

As the Indians do not take out the trees immediately, some of them may decompose, give back food to the earth and protect the ground from too much rain or sun. On the other hand, the Indians do not clear large areas, and within their small clearing, they sow different species. Their aim is to duplicate the variations of the natural tropical forest. They also occupy small areas (never more than two to five hectares) and never longer than three years. The Indians do not use fences, so, when they leave, the neighbouring forest returns and slowly the lost fertility also returns there.

The problems of some 400,000 Indians in Colombia are more or less the same as those of all Indians in South America. Without modern education, these '*Indios*' cannot fight for their rights. That Colombia has often tried to solve their problem is proven by the large volume *Legislacion Indigena Nacional* compiled by Adolfo Triana Antorveza and published by America Latina in 1980. It demonstrates that Colombia has been unable to do full justice to the Indians claiming land rights, with the result that the Indians have been threatened or bought by the *guerrilleros* who have been bribed by the mafiosi.

Of these 400,000 Indians, some 101,000 live in the Cauca; of these 58,000 are Paezes and their CRIC is the most powerful Indian organization linked up with the Coordinadora Nacional Indigena. There is also a recent legal project which was presented to the Congress of the Republic of Columbia in 1980 and speaks of the development of a fund for the furtherment of the Indians' timber and of other dispositions. Point 1 reads: 'The present law holds as its fundamental purpose to determine a political system in the state of Colombia which guarantees to its indigene population the conditions of life and the development compatible with the dignity of the human person.' This would have fulfilled Juan Quintin Lame's claim. This Paez Indian was born in 1880 and baptized a Catholic in Popayán. He professed his love for and his faith in the Virgin Mary, whom he venerated until his death in 1967. All his long life he fought for the rights of the Indians to their *resguardos* (reservations), their earth. He was elected to head the Cauca movement of the Indians of his time. During the last two years which he spent in prison he wrote a book published later under the title *In Defence of My Race*. Herein he writes: 'I am a defender in full light before God and man, who defends the tribes and partisans of my race of the earth Guanani; dead, disposed, analphabetic, abandoned, sad and lamenting for civilization.'

In Book 1, he writes also: 'The present book will serve as an horizon in the darkness for the indigene generations who sleep in the big camps which Divine Nature has eroded.' And on *page 14*: 'Nature is the book of God and of Science ... the thought of the smallest ant is the same as that of the condor and of the sons of men; because the ant, as it develops its wings, does not follow the same road as the others, because it climbs over the sand and beats its wings and it seems as it were defying eternity; because it feels itself potent; but, as it makes its way, it is attacked by its enemy, and in the same manner, error attacks all men.'

# Glossary:
## Spanish/Indian terms used

| | |
|---|---|
| *Ambil* | Extract of tobacco |
| *Bazooko* | By-product of cocaine |
| *Cabildes* | Chapters of Indians, elected by them |
| *Cacique* | Chief of the Witoto Indians |
| *Cepo* | A wooden board with holes for feet, used to torture prisoners |
| *Chacra* | A piece of land cleared by the Indians for their farms |
| *Cumare* | Natural cord plant |
| *Curaça* | Chief of the Tikuna Indians |
| *Danta* | Tapir, a tropical pig |
| *Encomienda* | Estate granted by Spanish kings |
| *Internado* | Mission boarding-school for Indian children |
| *Lideres* | Leaders |
| *Maloca* | Hut constructed by the Indians, usually for several families |
| *Minga guayuria* | Communal labour, usually for the construction of dwellings |
| *Narcomején* | The Eloria Noyesi larva, a grub that eats only coca leaves |

*Nofuico, La Chorrera* Waterfall
*Tatusino brasileño*    Cool drink with alcoholic content
*Uigobi*                Troubled waters

# Sources and Bibliography

*General Information*
*Colombia Amazonica* (Universidad Nacional de Colombia, Fondo para la protección del Medio Ambiente 'Jose Celestino Mutis', Fen-Colombia, Bogotá, first edition 1987), written by a group of the university's professors. Of especial interest are the following:

Gutierrez, Mario Mejia, '*La Amazonia colombiana, introducción a su historia natural*', particularly pp. 117, 121, 127 (deforestation)

Levya, Pablo, '*La Amazonia colombiana en perspectiva*', pp. 281, 284, 290, 293, 294 (general problems)

Reichel, Elizabeth Dussan de, '*Asentamientos prehispánicos en la Amazonia colombiana*', pp. 129, 131, 269

## 1. *Gold and Blood*

Chapman, C.E., *A History of Spain* (Macmillan, 1918), p. 221 etc

Parry, J.H., *The Spanish Seaborne Empire* (Hutchinson, 1966)

Unzueta, Antonio, OCD, '*La corona española responsable de la evangelización de América*', *Revista de Misiones* (Bogotá), 1987, no. 635/6

## 2. *Defence of a Mission*

Alcácer, Antonio, OFM cap., *El Convento del Socorro* (Ediciones Paz y Bien, Valencia, Spain, 1960)

Canyes, Mgr Marceliano E., *Geografia de la Comisaria Especial del Amazonas y Notas Historicas* (Prefectura Apostólica de Leticia, 1977)

Holguín, Alvaro Soto, 'Historia General de la Región Amazonica', *Revista Colombiana de Antropología* (Bogotá) 1975, pp. 33–59

Jover, Antonio, OCD, *Datos para la Historia de la Prefectura Apostólica de Leticia* (CEPROIAC, Leticia, 1985)

*Obra de los Misioneros Capuchinos Caquetá y Putumayo* (Prefectura Apostólica del Caquetá y Putumayo, Bogotá, 1912)

Moncanill, Gaspar Mignell, *Los Capuchinos 1906–33. Breve reseña de la obra de los misioneros Capuchinos en el Caquetá y Putumayo durante los años 1906–1933* (Impresa de la Cruz, Bogotá, 1912)

Vargas, Hectar Llanas, and Camacho, Roberto Pineda, *Etnahistoria del Gran Caquetá (Siglos XVI–XIX)* (Fundación de Investigaciones Arqueológicas Nacionales, Banco de la Républica, Bogotá, 1982)

3. *Hardenburg's Courage*

Casement, Consul Sir Roger, *The Blue Book* (1912/13 report), pp. 282, 284, 290, 292, 326, 328 etc, based upon Hardenburg, Ernst, *The Putumayo, the Devil's Paradise* (Truth, 1907/08)

Collier, Richard, *The River that God Forgot, the Story of the Amazon Rubber Boom* (Collins, 1968)

Correa, Carlos Larrabure (Director of Information of the Government of Peru in Europe), *Peru y Colombia* (published by same, Paris and Barcelona, 1913)

Guyot, Mireille, 'La Historia del Mar de Danta, El Caquetá, una fase de la evolución cultural en el Noroeste amazónico', *Journal de la Société des Américanistes* (Paris), 1979

Whiffen, Thomas, Captain W., *The North West Amazonas* (1915)

Yepez, Benjamin, Ch., *La Estatuaria Murui-Muinane* (Fundación de Investigaciones Arqueológicas

Nacionales, Banco de la República, Bogotá, 1982)

4. *The Caquetá*
Holguín, Alvaro Soto, op.cit., pp. 50–59
*Revista de Misiones* (Bogotá), 1977, Year LII, no. 578
Vargas, Llanos Hector, and Camacho, Pineda Roberto, op.cit.

5. *Dr Paul Rivet*
Gomez, Luis Duque (Director of the Colombian Institute of Anthropology), *Paul Rivet, Americanista y Colombianista* etc (Sociedad de Antropologia, Bogotá, 1958)
Reichel, Alicia Dussan de, '*Paul Rivet y su epoca*', *Correo de los Andes* (Bogotá), 1984, no. 26
Rivet, Paul, *Les origines de l'homme américain* (Mason, Paris, 1943–57)
———, *La conférence de réduction et de limitation des armements* etc (Geneva, 1932–4)
Soustelle, Georgette, *Paul Rivet, Médecin Militaire, Fondateur du Musée de l'Homme (1878–1958)* (Paris, 1976)

6. *La Chorrea – Nofuico*
Faerito, Rafael-Marciana Omi, *Nofuico 1933–1983*, with notes and commentary by Jaime Pujol Prats OCD (CEPROIAC, Leticia, 1983)
Sánchez, Cristobal Torralba OCD, '*50 años de NOFUICO – escuela internado La Chorrea, Misionero Capuchino desde 1941–1983*, *Revista de Misiones* (Bogotá), 1983, no. 578, pp. 288, 289 etc

7. *The Witotoes*
Castellvi, Marcelino de, OFM cap., *Manual de Investigaciones Linguisticas* (Imprenta Departamental, Pasto, 1934)
———, *Miscelanía CILEAC* (Sibundoy, 1952)
CILEAC: *Centro de Investigaciones Linguisticas y Etnograficas de la Amazonia Colombiana* (Sibundoy, 1940), vol. I, nos. 2–3

Gasché, Jürg, *L'habitat Witoto, progrès et tradition* (Neuchâtel, 1982) and *Journal des américanistes de Paris*, vol. 58 etc

Jover, Antonio, OFM cap., '*Amazonas, un magnifico Museo del Arte*', *Ethnia (organo del museo del indígeno)*, March 1981, vol. XI, no.60

Montclar, Fidel, OCD, '*Las Misiones Católicas en Colombia 1919, 1920, 1921*' (Sibundoy, 20 July 1920), *Revista Colombiana de Antropologia* (Bogotá, 1975), vol. XIX

Steward, Julian H., *The Handbook of American Indians*, Smithsonian Institute, Washington DC, USA, vol.III 'Witotan Tribes', pp. 43–9,

Yepez, B., op.cit.; also in *Colombia Amazonica* (Bogotá, 1987)

## 8. *The Mission's Work and Struggle*

The same references as for Chapter 7 and also:

Ruiz, Guillermo Lara, ed., *Castellvi, el Sabio y el Hombre. La obra antecedente de la Iglesia y de los Misioneros Capuchinos* (Academia Colombiana de Historia Eclesiástica, Medellin, 1986)

## 9. *Araracuara and the Andokes*

Camacho, Roberto Pineda, '*La Gente del Hacha, breve historia de la tecnologia según una tribu Amázonica*', *Revista Colombiana de Antropología* (Bogotá, 1975), vol. XVIII, pp. 441–77

Jover, Antonio, OFM cap., '*La Estación Misional de la Araracuara*', leaflet No.2 (CEPROIAC, Leticia, 1985)

Ortiz-Troncoso et G. Santos Vecino, *El Proyecto Colombo-Holandés Arqueocaribe* (Dutch-Colombian project, 1985)

## 10. *The Birth of Conservation*

Antorveza, Adolfo Triano, *Legislación Indígena Nacional* (Libreria y Editorial Americana Latina, Bogotá, 1980), pp. 80–82

Bonilla, V.D., *Siervos de Dios y Amos de Indios*, edited by the author (Bogotá, 1969)

Friedemann, Nina de and Arocha, Jaime, *Heredores del Jaguar y de la Anaconda* (Carlos Valencia, Bogotá, 1982), pp. 64, 65, 70, 72

*Revista de Misiones*: 'Amazonas 25 años' (Bogotá), July/August 1977, Year LII, no.578

Vidal, Ramón, '*Critica Historia a Siervos de Dios y Amos de Indios*', *Cultura Nariñense* (Pasto, 1970) No.25, pp. 37–120

11. *Two Congresses and one Ordination*

Antorveza, A.T., op.cit., pp. 321, 322

Idrobo, J.M., *II Simposio y Foro de Biologia Tropical Amazonica, Asociación Pro. Biologia Tropical del Foro en Florencia, Caquetá, 21 a 25 de Enero de 1969* (Pax, Bogotá, 1970)

12. *Differing Viewpoints*

Bergés, Yves-Guy, *La Lune est en Amazonie*, ed. Albin Michel (Paris, 1970)

Campuzano, Joaquim Molano, *La Amazonia, mentira y espera* (Ediciones de la Universidad de Bogotá, Jorge Tadeo Lozano, 1972)

Caycedo, German Castro, *Perdido en el Amazonas* (Plaza y Janes, Colombia Ltda, Bogotá, 1978/83)

Corry, Stephen, *Towards Indian Self-determination in Colombia* (Survival International, London, Document II, March 1976)

*Revista de Misiones*, No.578, as above

13. *The Tikunas*

Arbelaz, E. Perez, *Plantas útiles de Colombia*, ed. Camacho Roldan (Bogotá, 1956)

Cruz, Joaquim Paredes, *Colombia al Dia* (Ediciones CIMA, Bogotá, 1970)

Font, Juan A., OCD, partial translation of an article published in *Ethnia* (Bogotá) 1985, no.64

Whiffen, op.cit.

### 14. *Wasted Efforts*

Jover, Antonio, OFM cap., *Acción*, a Bible course for Indians (Leticia)

*Neue Zürcher Zeitung*: 'Wieder Indianermorde in Nordbrasilien' (Zürich), 15 April 1988

Pizano, Ernesto Samper, *Legalización de La Marijuana* (ANIF, Bogatá, 1980), p. 131

*Survival International News* (London), 1988, no.21, on March 1988 happenings


### 15. *Wicked Deeds and Other Complications*

Idrobo, J.M., 'El narcomején llega a Colombia', *El Espectador* (Bogotá), 2 December 1986

Osa, Verónica de, *Madre Laura, Mutter und Missionärin der Indianer* (EOS Verlag, Erzabtei St Ottilien, St Ottilien, West Germany)

*Time International*, cover-story, 25 February 1985


### 16. *Martyrdom*

*Revista de Misiones* (Bogotá), September 1987, no. 639

*Survival International News* (London), 1985, no.11 (Stephen Corry)


### 17. *What Future Is There for the Indians?*

*Acción 1987/88* (Leticia)

Arboleda, Diego Castrillón, *El Indio Quintin Lame* (Tercer Mundo, Colleción Tribuna Libre, Bogotá, 1973)

'Caja Agraria' 'Plan del Predio Putumayo' and *Campesino* (Bogotá, 1987)

Friedmann, Nina S. de, and Jaime Aroche, op.cit. pp. 72, 75, 76

# Index